JOURNEYS IN COMMUNITY-BASED RESEARCH

JOURNEYS IN COMMUNITY-BASED RESEARCH

EDITED BY
BONNIE JEFFERY
ISOBEL M. FINDLAY
DIANE MARTZ
AND LOUISE CLARKE

University of Regina Press

Copyright © 2014 University of Regina Press.

All rights reserved. No part of this work covered by the copyrights hereon may be reproduced or used in any form or by any means—graphic, electronic, or mechanical—without the prior written permission of the publisher. Any request for photocopying, recording, taping or placement in information storage and retrieval systems of any sort shall be directed in writing to Access Copyright.

Printed and bound in Canada at Friesens.

The text of this book is printed on 100% post-consumer recycled paper with earth-friendly vegetable-based inks.

Cover design: Duncan Campbell, University of Regina Press.
Text design: John van der Woude Designs.
Editor for the Press: Donna Grant, University of Regina Press.
Index: Patricia Furdek, Ottawa

Cover and interior images: Airphoto supplied courtesy of Saskatchewan Geospatial Imagery Collaborative.

Library and Archives Canada Cataloguing in Publication
Cataloguing in Publication (CIP) data available at the Library and Archives Canada web site: *www.collectionscanada.gc.ca* and at *www.uofrpress.ca/publications/journeys-in-community-based-research*

10   9   8   7   6   5   4   3   2   1

University of Regina Press, University of Regina
Regina, Saskatchewan, Canada, S4S 0A2
tel: (306) 585-4758 fax: (306) 585-4699
web: www.uofrpress.ca

U OF R PRESS

The University of Regina Press acknowledges the support of the Creative Industry Growth and Sustainability program, made possible through funding provided to the Saskatchewan Arts Board by the Government of Saskatchewan through the Ministry of Parks, Culture, and Sport. We also acknowledge the financial support of the Government of Canada through the Canada Book Fund for our publishing activities. We acknowledge the support of the Canada Council for the Arts for our publishing program.

CONTENTS

Acknowledgements vii
List of Abbreviations viii

Forewords
Jim Randall and Kate Waygood ix
Ron Labonte xiii

Introduction: Setting Out on Our Journeys xvii
Bonnie Jeffery and Louise Clarke

Section 1: Ethics of Community-Based Research

Working Together: Ethical Practice in Community-Engaged Research 3
Diane Martz and Juanita Bacsu

Talking to the "Healing Journey" Interviewers: Ethical Concerns and Dilemmas 15
Wendee Kubik, Mary Hampton, Darlene Juschka, Carrie Bourassa, and Bonnie Jeffery

The Ethics of Engagement: Learning with an Aboriginal Cooperative in Saskatchewan 29
Isobel M. Findlay, Clifford Ray, and Maria Basualdo

Section 2: Advocacy and Community-Based Research

Community-Based Research and Advocacy for Change: Critical Reflections about Delicate Processes of Inclusion/Exclusion 53
Gloria DeSantis

A Provocative Proposition: Linking Research, Education,
and Action in Saskatoon's Core Neighbourhoods   73
Mitch Diamantopoulos and Len Usiskin

Section 3: Impact of Community-Based Research

The CUISR-SRIC Collaboration: Toward
Community-Based Action Research?   91
Lori Ebbesen, Janice Victor, Louise Clarke, and Nicola Chopin

Community-Based Research through Community
Service Learning: Benefits and Challenges
for the Community and University   106
Nicola Chopin, Maria Basualdo, and David McDine

Tripartite Collaboration and Challenges: Reflecting on the
Research Process of a Participatory Program Evaluation   122
Hongxia Shan, Nazeem Muhajarine, and Kristjana Loptson

Standing Buffalo First Nation Youth:
Exploring Health and Well-Being   137
*Pammla Lusenga Petrucka, Roger Redman, Deanna Bickford,
Sandra Bassendowski, Andrea Redman, Leanne Yuzicappi,
Bev McBeth, Logan Bird, and Carrie Bourassa*

Knowledge Translation Strategies in Community-Based
Research: Our Decision-Maker-Based Approach   149
Fleur Macqueen Smith, Nazeem Muhajarine, and Sue Delanoy

Conclusion   163
Isobel M. Findlay and Diane Martz

Contributors   **171**
Index   **177**

ACKNOWLEDGEMENTS

The journey to bring this manuscript to publication has been a long one and we gratefully acknowledge the perseverance and commitment of each of the contributors to this volume. Our ability to produce this collective work by bringing together a range of authors to document their experiences was supported by the Saskatchewan Population Health and Evaluation Research Unit (SPHERU) at the Universities of Regina and Saskatchewan and the Community-University Institute for Social Research (CUISR) at the University of Saskatchewan.

We would like to acknowledge the financial support provided by the Saskatchewan Health Research Foundation (for support of SPHERU), the University of Saskatchewan's Publications Fund, the Saskatoon Regional Intersectoral Committee (for support of Ebbesen, Victor, Clarke, and Chopin chapter), and the Social Sciences and Humanities Research Council of Canada for support of Linking, Learning, Leveraging: Social Enterprises, Knowledgeable Economies, and Sustainable Communities, Regional Node of the Social Economy Suite (chapters by Findlay, Ray, and Basualdo and Diamantopoulos and Usiskin).

Thank you to the two anonymous reviewers for their constructive comments and suggestions and to Dallas Harrison for his considerable expertise in editing the manuscript for submission to the Press. We are also grateful to the University of Regina Press, especially to Bruce Walsh, executive director, and Donna Grant, senior editor, for their generous support and guidance.

List of Abbreviations

CBPAR	community-based participatory action research
CBPER	community-based participatory evaluation research
CBPR	community-based participatory research
CBR	community-based research
CIHR	Canadian Institutes of Health Research
CPHI	Canadian Population Health Initiative
CSL	community service learning
CUISR	Community-University Institute for Social Research
CURA	Community-University Research Alliances
ECDU	Early Childhood Development Unit
EDI	Early Development Instrument
ESL	English as a Second Language
EWG	Evaluation Working Group
GIS	geographic information systems
KT	knowledge translation
NGO	non-governmental organizations
NPO	non-profit organizations
NSTAC	Northern Saskatchewan Trappers Association Cooperative
RCAP	Royal Commission on Aboriginal Peoples
RESOLVE	Research and Education for Solutions to Violence and Abuse
RIC	Regional Intersectoral Committee
RSAT	RIC Self Assessment Tool
SPHERU	Saskatchewan Population Health and Evaluation Research Unit
SRIC	Saskatoon Regional Intersectoral Committee
SSHRC	Social Sciences and Humanities Research Council
TCPS	Tri-Council Policy Statement
UEY	Understanding the Early Years

FOREWORD
Jim Randall and Kate Waygood

It is gratifying to see the Community-University Institute for Social Research (CUISR) celebrating its tenth anniversary. CUISR's rich history can be traced to 1997, when a core group from community-based organizations, civic departments, the regional health authority, and academics from Geography, Community Health and Epidemiology, and Native Studies formed the Ad Hoc Committee for Quality of Life. The initial aim to draft a proposal to both the university administration and the community organizations (e.g., City of Saskatoon, the Saskatoon Health District) coincided with a call for proposals for a new Social Sciences and Humanities Research Council of Canada (SSHRC) funding program called the Community-University Research Alliances (CURA).

The SSHRC initiative was tailor-made for our group. All partners wanted to understand better how to build and sustain a healthy community. Each sector was to commit to a three-year initiative in the belief that the partners could better develop sound social policy and resolve pressing social problems by working in concert. The proposed alliance presented challenges, including how a group representing such diverse organizations, each with its own agenda and goals, could sustain a truly meaningful partnership with an equitable system of governance.

Why did community organizations want to get involved with CUISR? (from the perspective of Kate Waygood)

As a city councillor (1979–2003), I used my background in geography and community activism to guide my decision to invest time and energy in this initiative. I held to the belief that those whose lives were affected by government policies should have a greater voice in how those policies were developed. Saskatoon City Council reviewed many requests for additional funding for existing or new programs. Too often council had turned down requests because the proponents, whether administrators or the public, needed better and more reliable evidence to support the request for funding. Decision makers had to believe that in the end they would either save money or obtain better outcomes by supporting the request. Without authoritative, long-term studies, this was often hard to prove in fields such as social services. Many elected officials and staff in government, as well as those from community-based organizations and the community at large, were looking for ways to strengthen arguments to support more comprehensive and coordinated public policies. The legitimacy associated with an organization such as CUISR provided that vehicle.

Why did academics and the university want to get involved with CUISR? (from the perspective of Jim Randall)

A core objective of the group of academics committed to applied, participatory research was to produce more direct and immediate changes in community quality of life. However, with some notable exceptions, these research projects were limited in scope and duration, and did not receive systematic support from the university or government. These academics wanted the university to take a more active, coordinated, and long-term role in supporting community-engaged scholarship and translating the knowledge into community action.

The university supported the initiative's potential to build social research capacity among faculty and students and to help realize the university's community service mandate. However, support also had a more practical dimension. The 1999 CURA grant of $600,000 over three years was the largest SSHRC grant the university had ever received. This funding (and a further $400,000 three years later) allowed the university to leverage additional federal money for indirect research operating costs and counted towards the number of Canada Research Chairs awarded to the university.

Initial Premises and Structures of the Institute

CUISR's success was based on a vision of the community developed collaboratively and was informed by such foundational principles as partner equality and honesty. Over countless meetings, the diverse partners began to trust each other, to value their respective knowledge bases, and to trust that the needs of all would be addressed. Without time invested in reflecting on, recording, and applying lessons learned from community-university research initiatives, partnerships could have faltered. This sense of trust was important because the rules and structures of the research-funding body and the university did little to promote equality in decision making. For example, all of the funds had to be held and managed at the university, and the university co-director theoretically had sole signing authority regardless of the protocols put in place by the institute.

Although CUISR's original mission was "to serve as a focal point for community-based research and to integrate the various social research needs and experiential knowledge of the community-based organizations with the technical expertise available at the University," it quickly became apparent that this construction of a community-university research partnership was presumptuous and simplistic. Not only did community advocates and participants bring a rich body of expertise to the table, but also university academics were not just impartial observers disconnected from the social life of the community.

The CUISR governance model was intended to foster continued cooperation and trust while minimizing administrative bureaucracy. It was a partnership of equals: lived experiences and in-kind contributions matched degrees and dollars. The Management Committee of eight, four each from the university and the community included the community and university co-directors as well as the community and academic directors for each of the three research modules. The modules developed their own research agendas within the overall institute mandate and goals. This group of eight vetted all substantive budget decisions and adjudicated all proposals for research assistance. A CUISR research coordinator appointed to liaise between the university researcher and the host community-based organizations kept the lines of communication open and clear throughout the research initiatives.

CUISR was equally committed to disseminating research results in ways useful and meaningful to those ultimately affected by the policies and programs in addition to the usual peer-reviewed outlets. Research results were spread at "brown bag lunches," in newspaper advertisements, at public policy forums and town halls, in skits by local theatre troupes, and even in an insert in the local Saskatoon Star-Phoenix newspaper. At a time when few were connected to the Internet, this insert reached readers in 76,000 homes.

Over ten years later, policy-makers, decision makers, the public, and the participants are still learning from CUISR's initiatives. Better programs and policies are being approved, students and faculty are learning first-hand of the social challenges facing Saskatoon, the capacity and legitimacy of community-based organizations are being enhanced, and research is making a difference in the lives of the less advantaged. Each step in this ten-year path has provided lessons for sustainable and successful community-university initiatives. New partners and investments continue to reap benefits for all. Indeed, the whole has proven to be greater than the sum of the parts.

FOREWORD
Ron Labonte

My tenure as the founding director of the Saskatchewan Population Health and Evaluation Research Unit (SPHERU) coincided with a personal transition and a professional opportunity. My personal transition was acceptance of a full-time position in a university setting after two and a half decades of work in government and international consulting. At the same time, I was presented with a professional opportunity to enter into a partnership with the Community-University Institute for Social Research (CUISR). Embracing the idea of a partnership with CUISR when I took on the SPHERU directorship seemed only natural. CUISR and SPHERU, in the early years, were similar to conjoined twins. Several of SPHERU's researchers were also affiliated with CUISR; community-based research was one of our several population health research interests and one shared with other CUISR researchers. Building a partnership based on this foundation was a logical step in the evolution of both organizations.

Jim and Kate honestly comment above on some of the tensions that can beset the community-university relationship. They also note that "we" (the academy) are also "they" (the community), with the blurry line between the two that comes into focus only when their differing knowledge premises, subject positions, and potential roles in creating healthier communities are recognized and respected. However, "we" inside the academy are also communities with our own dynamics.

This became a "practice what you preach" moment for me. My experience of working outside the academy guided me to avoid competition over ownership of new initiatives. One never puts an organization in the centre and then configures other partners around it. One always puts the problem in the centre and then asks, "Who is best to orbit it and with what contribution?"

Having spent the first twenty-plus years of my public health career more firmly rooted in the community/government side than in the university side, I was impressed by the integrity with which community and university partners in CUISR struggled to get the relationship right. I was also struck by some of the compromises and quick learning that we (and I speak now of the researcher side of the teeter-totter) had to make in the early years.

One such example was when the university liaison for the Saskatoon Star-Phoenix announced that the newspaper would run a special issue based on a survey that SPHERU and CUISR wanted to do about quality of life in Saskatoon. Although the paper was willing to front many of the costs and would help to manage the process, the wrinkle was that the special issue was due to hit the streets approximately three or four months after the offer was made. There was an audible gasp among the startled researchers around the table. They understood that it would take months just to develop the survey instrument and validate it, much less undertake the survey and do the analysis. Maybe in a year or so it would be possible. This reaction attests to the well-known temporal disjuncture between academia and commercial or community interests. One of the clichés about universities is that they do move but at about the speed of glaciers—and that's pre-climate-change speed.

The offer was too good to refuse, and the challenge galvanized our group with a topic, a problem, and a deliverable demanding that all of us work together. To "validate" the instrument, cobbled together with rigour as well as rapidity, we doubled up with a community group meeting at which over seventy representatives were collectively administered the survey, reflecting in table groups on what was great and should stay in, what was okay but could be cut, what was not in but really should be, and what did not make sense and needed more thought. Voila! Community buy-in, a better survey, and a bit of fun to boot.

Then there was the "So what?" meeting with community groups and policy decision makers after the survey was completed and analyzed. We had so much great data and even some regressions. Just think of all the charts that we could render into mind-numbing PowerPoints!

Instead, and taking a cue from common-sense communication 101, the findings were distilled to three headline messages. The data were presented as briefly as possible, combining survey results with interview and focus group narratives. An Aboriginal theatre troupe dramatized each of these key messages in short, powerful vignettes, providing an emotive connection to the data. A review of what community organizations were already doing to act on these headline messages was presented, leading to a gap analysis of what more needed to be done. Time might be rose-tinting my memory specs, but I recall this event as one of the finest examples I have experienced of successfully bridging research findings and community perspectives.

But this led inevitably to the next question: "Now what?"

It is one thing for a community-university research partnership respectfully to combine different forms of knowing and collectively to identify which programs and policies need doing or changing. It is quite another to commit to the work to make the changes. Among some of the researchers, who were then busying themselves with the required write-ups for publication on which their careers (and future value-added to community-university partnerships) depended, the sentiment went something like this: "Well, community groups have the data, and they have the ideas and the proposals. What more can we give them?"

If community groups sometimes do not appreciate the "grant and publish" treadmill of academic researchers, researchers sometimes do not appreciate that community groups are consumed with existing programs, projects, and work demands, in which it is often difficult to act upon new knowledge (however important or useful). If one great collaborative lesson was generated in CUISR's early years, it was that collaboratively generated knowledge does not automatically translate into action. Development of resources to move the new knowledge forward is as important as resources to develop the new knowledge in the first place.

The quality-of-life survey has since gone through another couple of waves. Dozens of research projects, small and large, have commenced, been completed, and moved forward through CUISR's ten years. But neither communities and organizations nor universities and their researchers are static. Some of CUISR's original partners, in both the community and the university, have moved to different places.

The test of any good partnership lies in its ability to shed old individuals while absorbing new ones. It is one that SPHERU has weathered with gold-star kudos, and I am proud to have played a small role in its inception and early years, and I am tickled prairie pink (or whatever hue of colour the reader prefers) to see some of the fruits of SPHERU's endeavours in this wonderful new collection.

INTRODUCTION
SETTING OUT ON OUR JOURNEYS

Bonnie Jeffery and Louise Clarke

The purpose of this volume is to recount and celebrate three different journeys. First are the journeys of the authors, who have navigated the challenging and ultimately rewarding waters of community-based research (CBR). All of the authors in this volume are or have been affiliated with the Saskatchewan Population Health and Evaluation Research Unit (SPHERU) and Community-University Institute for Social Research (CUISR). With this volume we also celebrate the journeys over the past decade of our two research institutes to fulfill the founders' vision of genuine community-university research partnerships in the service of better lives for vulnerable people and groups. And third is our own journey as editors of this collection, seeking to build our partnerships and practice of community-based research (CBR) in the process of working with the authors.

We have organized the volume to highlight examples of the successes and challenges of CBR across a range of projects in three areas of interest: ethical issues in CBR, issues that arise in CBR projects with an advocacy focus, and the impact of CBR projects. The first section, "Ethics of Community-Based Research," presents chapters that address ethical issues faced in the development of university-community partnerships and the engagement of communities, such as power imbalances, understanding and respecting cultural diversity, using culturally competent practices, participation, and community capacity building. The chapters addressing "Advocacy and Community-Based Research" discuss specific advocacy strategies or methods that have been employed, such as community meetings, community advisory groups, policy roundtables, and community-university partnerships. And, finally, case studies that highlight the "Impact of Community-Based Research" include examination of a specific CBR initiative that has led to an identifiable change in policy, program, or capacity development in reducing various inequalities.

That SPHERU and CUISR should come together to produce this book is not surprising given our parallel and often overlapping journeys. As described by Randall and Waygood and Labonte in their forewords to this volume, some of the same people were engaged in discussions about community-university partnerships to advance health and quality of life in 1999. All were deeply influenced by the specific location, Saskatchewan, and its socio-economic, geographic, and cultural contexts. All were committed to adopting multiple perspectives and CBR principles, but each organization took a slightly different path to reach its goals. We will provide brief overviews of SPHERU and CUISR and then map the terrain of CBR that we share.

SPHERU

SPHERU was established jointly, in 1999, by the Universities of Saskatchewan and Regina as an interdisciplinary research unit committed to the promotion of health equity by understanding and addressing population health disparities through policy-relevant research. Researchers at SPHERU come from a variety of academic backgrounds, including geography, political science, anthropology, epidemiology, social work, economics, nursing, nutrition, and history. Although there is no unifying theory of population health per se (Coburn et al., 2003; Kindig & Stoddart, 2003; Labonte et al., 2002), these researchers draw on discipline-specific theories of health determining conditions related to social class, gender, culture, society, place, and time. The researchers actively engage with communities and policy-makers to accomplish the goals of the unit, which include building on the existing expertise, knowledge, and capacity of all research partners and exchanging research knowledge with communities and policy-makers through ongoing engagement (McIntosh, Jeffery, & Muhajarine, 2010). At the foundation of SPHERU's mission is a critical population health approach to research that explicitly addresses health disparities and is "shaped by a consideration of the interests of those who face the greatest burden of disease" (McIntosh et al., 2010, p. xv).

SPHERU's research focus is primarily concentrated in three broad theme areas with attention to investigator (academic or community) initiated projects and evaluation studies that fall within these themes. Our research work in the area of Northern and Aboriginal Health seeks to develop culturally relevant health frameworks and examines the role that culture plays as a health determinant, the Healthy Children research program considers how contexts shape children's health and development outcomes, and our Rural Health research focuses on the impacts of socio-economic determinants of health in rural populations. Foundational to our research work is the commitment to a collaborative approach to population health research, working with communities and policy-makers by

- mobilizing and building on the expertise of our researchers, students, trainees, and research partners;
- ensuring that our research questions and results are relevant to improving the health of Saskatchewan residents; and
- exchanging our research knowledge through engagement with communities and policy-makers.

The majority of our research incorporates community-based principles, with projects initiated by either university or community researchers in which the work is supported through external grants (e.g. Canadian Institutes of Health Research) and contracts and some modest support for administrative staff from both universities. SPHERU has also received team grant funding from the Saskatchewan Health Research Foundation to support ongoing work in our identified research areas. Our governance structure involves active participation of all researchers and staff and is formally guided by a memorandum of understanding signed by the University of Regina and the University of Saskatchewan that outlines the cooperative agreement to support our work. Institutional support is provided by the Management Advisory Group, whose four members are appointed by the vice-president research at each university to provide consultation and support to the director and members of SPHERU.

CUISR

A group of community and university people committed to improving the quality of life in Saskatoon came together in the late 1990s and successfully applied to SSHRC in 1999 for a Community-University Research Alliance (CURA) grant to establish CUISR and conduct research. The institute was formally recognized by the University of Saskatchewan as a university-wide (interdisciplinary) centre in 2000, with the mission to "facilitate partnerships between the university and the community in order to engage in relevant social research that supports a deeper understanding of our communities and that reveals opportunities for improving our quality of life." To fulfill the mission of facilitating partnerships, the institute is uniquely governed at all levels—board, co-directors, thematic areas, and project teams—by half community and half university representatives working cooperatively to reach mutually beneficial decisions through consensus. Almost all projects are initiated as the result of an expressed need from the community.

Faculty members from many disciplines are not specifically appointed to CUISR but rather choose to serve in various capacities. As a result, the institute does not offer any formal program or courses, and faculty do not receive time releases from teaching to do CUISR research (unless a faculty member is the principal investigator of a major research grant). Notwithstanding this limitation, faculty together with community partners, institute staff, and graduate students have created an impressive research output over the years in areas including housing and homelessness, poverty reduction, food security, individualized funding policies for people with disabilities, community economic development, cooperatives, and many others. In addition to in-house reports and academic publications, CUISR is committed to presenting results to the community through brown bag lunches and community forums where members of the public are invited to review findings and propose new work. This work has been supported by three sources of funding. The first is research grants from SSHRC, including two major CURA grants for research on quality of life in Saskatoon and area and on the social economy in Saskatchewan (the latter was done in partnership with the Centre for Study of Co-operatives at the University of Saskatchewan). CUISR received another grant in 2003 to host the first CU Expo in Canada on community-university endeavours. The second source of revenue is grants and in-kind assistance from our key partners over the years: the University, the Saskatoon Health Region, the City of Saskatoon, and the United Way of Saskatoon and Area. Starting in 2007, we began conducting research projects for government agencies and non-governmental organizations (NGOs) on a fee-for-service basis. Projects must be in line with the institute's strategic directions and adhere to CBR principles as much as possible, and the results must be in the public domain. Over the past decade, CUISR has maintained and even deepened its commitment to a highly participatory form of CBR.

Mapping the Terrain of CBR

In this section we present an overview of CBR foundations and recent developments and of the three areas that we invited authors to address in their CBR case studies and experiences— ethics, advocacy and impacts— recognizing that there are theoretical and practical links among them. We highlight the importance of relationships in the research partnership that recognize the central values of power, voice, and control throughout all aspects of the partnership and process along with our reflections on issues specific to the university-community relationship that can influence the ethical practice of CBR. We then reflect on the advocacy focus of this type of research through a discussion of participatory methods that are the basis of genuine CBR and that address the contribution of methodologies that are participatory, emancipatory, and decolonizing. The final section highlights some of the key issues that arise when we discuss the impacts of CBR and, in particular, the meaning of "evidence" and its influence on the application of findings for change.

Foundations and Recent Developments of CBR

CBR stems from both action research and participatory research (Flicker, Savan, McGrath, Kolenda, & Mildenberger, 2007). Action research incorporates active participation of all parties in decision making through a cyclical process that occurs throughout the project (Carlisle & Cropper, 2009; Flicker et al., 2007), is open to change, and accepts the unpredictability of outcomes (Carlisle & Cropper, 2009). Dialogue, reflection, consciousness raising, mutual skill development, and shared power based on fairness and equity (Rogge & Rocha, 2005) describe participatory research. Together these two traditions focus on problem solving and change in the community.

More recently, CBR has been discussed as one approach within the overarching concept of the scholarship of engagement. Within universities, the scholarship of engagement, sometimes defined as a movement (Sandmann, 2008), refers to a greater commitment to the importance of civic engagement with the production of knowledge (Barker, 2004) through connecting faculty expertise to public issues in collaboration with communities (Stanton, 2008). Engaged partnerships with communities can include teaching and research components through specific practices such as service learning, participatory research, and CBR.

We recognize that there is no one accepted definition of CBR and that, in fact, it can be thought of as a continuum. Although Savan and Sider (2003) note that most forms of CBR involve a collaborative approach in which decision making is shared, this is not always the case. There are different levels of participation along a "continuum of control," which ranges from the academic researchers at one end to the community at the other (Wallerstein & Duran, 2006). Others use the term "community-engaged research," which refers to a continuum of research along which the roles of community and academic research partners can vary (Ross et al., 2010). These broad definitions, however, run the risk of identifying research as community based when, in fact, it does not conform to the basic tenets of collaboration and partnership in CBR.

Our approach to community-based research, exemplified by many of the case studies that follow, is based on a commitment to research that is action oriented and directed at reducing inequities through recognition of the importance of values such as self-determination, protection of confidentiality, equal distribution of resources, recognition of power issues, and promotion of cultural diversity.

Ethics of CBR

At the heart of any discussion of CBR is the commitment to a value stance that explicitly supports the need for research to contribute to change or the "social imperative" as the basis of any CBR initiative (Savan & Sider, 2003). The concept of ethics, then, is central to how community-university partner relationships are developed and sustained throughout the research process. In this brief overview, we focus on two issues that bear on ethical CBR practice: factors that can affect the university-community relationship, and ongoing tension about the place of this type of research within university institutions.

Aspects of time and funding highlight the sometimes different worlds occupied by the university and community partners, both having the potential to result in power imbalances in the research relationship. Time constraints faced by both partners (Ibáñez-Carrasco & Riaño-Alcalá, 2011; Tyron et al., 2008) and sometimes frustration over delayed results can impact the outcome of CBR (Rogge & Rocha, 2005). Increasingly, universities and research funding agencies, such as the Canadian Institutes of Health Research (CIHR), are valuing and supporting university-community research partnerships (Savan, 2004). Although this is a positive development, access to funding and other resources still resides primarily with academic researchers, thus limiting the ability genuinely to share control of important aspects of the research process. Roche (2008, p. 9) argues that "New opportunities for CBR funding, while promising, still situate the responsibility (and control) of research projects with academically affiliated investigators. With access to institutional supports for research somewhat limited in the community sector, including the lack of options for ethical governance and accountability, power imbalances remain firmly intact."

The University of Saskatchewan has explicitly included community-university partnerships as a key focus starting with its *Foundational Document on Outreach and Engagement: Linking with Communities for Discovery and Learning*, in which both SPHERU and CUISR were mentioned as exemplars (2006, pp. 17–20). Although engagement is a core commitment in current University planning documents (University of Saskatchewan, 2008, 2011, 2012a, 2012b), CBR has not been the major focus of attention. The 2010 Task Force on Community-Based Research coordinated by the Office of the Vice-President Research is acknowledged in the report *Engaging with external partners* (University of Saskatchewan, 2012b) and its recommendations to take advantage of synergies and to enhance the University's impact on community-university partnerships of all types. Outreach and engagement activities now part of the Office of Vice-President Advancement and Community Engagement have funding commitments and an office of Community Engagement at Station 20 West to support ongoing research, graduate and undergraduate student research, community service learning, Community Engaged Scholar Discussion Group, an Engaged Scholar Day, innovative teaching, a planned journal, and other engagement efforts. All of these initiatives open up institutional and other space for SPHERU and CUISR.

The University of Regina has recently approved a strategic research plan that identifies CBR as a subtheme in the signature theme area of knowledge creation and discovery (University of Regina, 2011). Although the research plan supports the "explicit engagement of community stakeholders in research endeavours whenever and wherever possible" (p. 4), there is little in the strategies that would suggest a commitment to the necessary institutional supports required for successful CBR. These broad initiatives at both the University of Saskatchewan and the University of Regina do not explicitly address expectations of faculty and the structures (tenure, promotion) within the institutions that ultimately can impede successful and sustained involvement in CBR. Several authors note that changes in how scholarship is defined along with an expanded view that would open a place for products of CBR are needed to advance this type of research within the academy (Calleson, Jordan, & Seifer, 2005; Roche, 2008).

Advocacy and CBR

A fundamental principle of CBR is the explicit focus on change as a result of the research conducted by the university-community partnership. Claims regarding whose knowledge is recognized and how it is expressed and/or validated are central to the advocacy and social change focus of CBR. Successful CBR, then, involves attention to the philosophical foundations of this approach as well as appropriate methods (Ibáñez-Carrasco & Riaño-Alcalá, 2011). Roche (2008) describes participatory methods as the "cornerstone" of CBR, and others argue that the most appropriate research methods in CBR must "go beyond the 'mere involvement' of those whose experiences are being researched to allow for their 'responsible agency' in the production of knowledge" (Salmon, 2007, p. 983).

Roche (2008) argues that CBR can make a contribution to innovation in research methods (e.g., peer research, arts-based methods) but tempers this assertion with a cautionary note that those committed to CBR must begin to articulate and critically appraise what constitutes participatory methodologies. This involves assessing whether these methodologies (and methods) are adequate in genuinely including community knowledge and voice in the research process. As noted by Wallerstein and Duran (2006, p. 315), we must "unpack the role of power and privilege in the research relationship" by critically assessing how knowledge is produced and validated.

Impacts of CBR

The third theme arising out of our discussions of CBR was impacts of the research. Roche (2008) highlights some of the dilemmas inherent in CBR making meaningful impacts or, more precisely, the differing perspectives on impacts. Scholars sometimes face a conflict between how the academy will judge impacts (e.g., validity, citations) and how the community partner will judge impacts (e.g., telling its story, actionable measures that reduce the problems confronting it). For policy-makers, the evidence of change must be compelling, which means that it needs to be linked to issues, data, and mechanisms of social change beyond the local sphere. This might not satisfy either the community or the scholar. Moreover, impacts can be difficult to discern, let alone quantify, especially within the relatively short time spans of a research project.

As a result, CBR needs to be assessed in terms of how well the research partners balance the various interests—rigour, relevance, action. The research might show direct impacts of interventions on policies or programs, but there might also be a case for indirect effects. The research might increase the capacity of citizens to participate in various stages of the research process or even the willingness of skeptics to become engaged (e.g., citizens who feel "researched to death" or bureaucrats concerned that the research will be too academic simply because it references social theory and other studies in different contexts). Last, and certainly not least, the research can contribute "new" knowledge to the academy, such as Indigenous knowledge, and innovative research methods, such as using peer researchers or the arts to tell stories and effect social change. Underpinning successful impacts are the willingness and ability of all participants to engage in critical reflexivity and open dialogue in the CBR process. Thus, learning key lessons about building community-university partnerships and doing mutually beneficial research is also a valuable impact.

We indicated in the first paragraph that the third journey we celebrate is our own as editors of this volume. We have enjoyed the cross-disciplinary dialogue and the resulting realization of how much we share in CBR principles and objectives. We have also (re)learned the difficulties of truly engaging our community partners in this aspect of the work due to several imbalances, notably time and money. Community partners rarely have the time or interest to write up the results of the research. A workshop to discuss first drafts of the chapters unfortunately did not draw many of our community partners—busy people who cannot usually spare a day for talking and writing—but did create a wonderful learning opportunity through open dialogue and critique for all who did attend. We did commit ourselves to making the material accessible to community participants and to students doing community service learning. Another imbalance is that academics have greater access to the resources to edit and publish a book than do most of our community partners.

Journeys in CBR

As previously mentioned, this volume celebrates ten years of community-based research conducted by SPHERU and CUISR by highlighting challenges, successes, and lessons learned on nine CBR journeys. The first section of this volume presents chapters that focus on ethical issues specific to this type of research, with attention paid to the roles and responsibilities shared between university and community partners. The first chapter, "Working Together: Ethical Practice in Community-Engaged Research," by Martz and Bacsu, summarizes findings from a qualitative study in which academic and community partners were interviewed on their experiences with ethical issues in a CBR project. The views of these participants highlight the need for continual negotiation on ethical issues through transparent and open communication.

The chapter by Kubik and her colleagues, "Talking to the 'Healing Journey' Interviewers: Ethical Concerns and Dilemmas," frames the discussion of ethics from the perspectives of the university- and community-based interviewers who collected information over the five years of this longitudinal study. They highlight the paramount issues of safety and confidentiality in conducting research among a vulnerable population of women who have been abused by their intimate partners. The original contribution of this chapter is their focus on the experiences of the interviewers themselves and the ethical challenges that they experienced during data collection. The importance of providing training for the interviewers is key to addressing these issues, but the authors acknowledge that efforts to support the interviewers through "debriefing" opportunities are best accomplished when these opportunities are formalized.

Findlay, Ray, and Basualdo focus on the process and ethics of engagement in discussing their multi-year research partnership between CUISR and the Northern Saskatchewan Trappers Association Cooperative in their chapter, "The Ethics of Engagement: Learning with an Aboriginal Cooperative in Saskatchewan." Relationships are fundamental to ethical CBR, and engaging with communities is a challenging, complex, and iterative process sustained by continual reflection on who we are as researchers. Key to successful engagement is commitment to decolonizing research that supports a process of dispersing authority and sharing power throughout the research process and that honours community wisdom. These are essential aspects of ethical engagement but are nonetheless challenging to implement. The discussion of the "ethics of place," which includes both community and university, is critical to successful and ethical engagement with communities. The question that we are left with is whether the university as a "place," along with its own cultural traditions, is fully supportive of embracing CBR as a valued scholarly activity.

The second section presents two chapters that discuss ethical, philosophical, and practical issues that can arise when CBR has a clear advocacy focus. DeSantis discusses a project in which urban residents addressed issues of access to the social service system and integrates her experience as a community activist and practitioner with her role as an academic researcher to reflect on the importance of addressing issues of power and voice through inclusive processes. In her chapter, "Community-Based Research and Advocacy for Change: Critical Reflections about Delicate Processes of Inclusion/Exclusion," DeSantis discusses how, through a CBR approach, a Steering Group was developed as a strategy for being inclusive of community members during the research process. This example highlights the fundamental issue that many researchers face when conducting CBR in determining the definition of the community with which they are working. The success or limitation of this strategy is influenced by understanding the complexities of participation in the research by defining who is involved and when and how they are involved. Participation is clearly linked to concepts of power and voice in CBR; there are different levels of power and varying abilities to exercise power during the research process.

In "A Provocative Proposition: Linking Research, Education, and Action in Saskatoon's Core Neighbourhoods," Diamantopoulos and Usiskin also consider how we define the community that we are working with in CBR. Their project clearly reflects the community development roots of CBR and the recent trend to use a critical theory lens. This chapter challenges us to acknowledge the conflicts created by class, race, and gender between core and suburban neighbourhoods, within core neighbourhoods, and between community residents and the university. The authors also show how critical theory and reflexivity by scholars can help to overcome some of the barriers between community and university by making the research project democratic and public, thereby enhancing effectiveness as advocates for the marginalized.

The final section of the volume includes several chapters that focus on the *impact of community-based research* that has addressed or reduced various inequalities through change in policy, program, or capacity development. Ebbesen, Victor, Clarke, and Chopin—in their chapter "The CUISR-SRIC Collaboration: Toward Community-Based Action Research?"—discuss the concept of "alignment" in their participatory evaluation project with the Saskatoon Regional Intersectoral Committee (SRIC) and the strategy of developing an Evaluation Working Group as an intermediary to support the university-community partnership. The key learning from this chapter is how a participatory evaluation process with a relatively powerful community partner, SRIC, can develop, which then helps them contribute more actively than in the past to social change. The community service learning program described by Chopin, Basualdo, and McDine highlights the impact that conducting CBR can have on students through their participation in a community-based project. Their chapter, "Community-Based Research through Community Service Learning: Benefits and Challenges for the Community and University," demonstrates the value to students of introducing CBR while they are trainees.

Overall, we often talk about CBR as a partnership between the academic and community (non-academic) sectors. Shan, Muhajarine, and Loptson expand our understanding of partnerships by discussing strategies that they employed with a tripartite partnership in the evaluation of *KidsFirst*, an early childhood intervention program. They discuss the challenges of sustaining a successful tripartite partnership among academic researchers, community partners, and government partners in "Tripartite Collaboration and Challenges: Reflecting on the Research Process of a Participatory Program Evaluation." Key to sustaining these relationships is recognizing the power differentials and different standpoints that exist among partners. Regular and frequent communication on co-learning and an open space for dialogue to discuss differences are important strategies.

Petrucka and her university and community colleagues address the impacts of CBR on the youth of a First Nations community in southern Saskatchewan. In "Standing Buffalo First Nation Youth: Exploring Health and Well-Being," they outline the impacts on both participants and the community of a project related to wellness challenges among Aboriginal youth. They highlight the necessity of being both sensitive to culture and flexible throughout the research process in order to adapt to community-driven research questions and methods and thus provide an example of innovative, arts-based research methods that have come out of a CBR project.

Macqueen Smith, Muhajarine, and Delanoy, in their chapter "Knowledge Translation Strategies in Community-Based Research: Our Decision-Maker-Based Approach," discuss the program and policy impacts of the Understanding the Early Years study, a seven-year university-community research partnership. They focus on the role of decision makers on the research team and how their involvement facilitated effective knowledge translation strategies for the project. They point to the implementation, based on the research findings, of major initiatives in literacy programs and literacy-enhanced kindergarten programs.

Our hope is that this book, directed primarily at community and university trainees and researchers new to CBR, will contribute to the theory and practice of an approach to research that has much to offer in terms of change at individual, community, and societal levels. Notwithstanding the many challenges of CBR, this is an exciting time to be part of its developing landscape as new funding and support opportunities become available to conduct this research. The chapters that follow address many of the issues outlined in this introduction, and we anticipate that these experiences will provide helpful guidance to others as they embark on their own journeys in CBR.

Acknowledgements
We wish to acknowledge SPHERU, CUISR, and the University of Saskatchewan Publications Fund for providing the necessary support for completion of this volume.

References
Barker, D. (2004). The scholarship of engagement: A taxonomy of five emerging practices. *Journal of Higher Education and Engagement*, 9(2), 123–137.
Calleson, D., Jordan, C., & Seifer, S. D. (2005). Community-engaged scholarship: Is faculty work in communities a true academic enterprise? *Academic Medicine*, 80(4), 317–321.
Carlisle, S., & Cropper, S. (2009). Investing in lay researchers for community-based health action research: Implications for research, policy, and practice. *Critical Public Health*, 19(1), 59–70.

Coburn, D., Denny, K., Mykhalovskiy, E., McDonough, P., Robertson, A., & Love, R. (2003). Population health in Canada: A brief critique. *American Journal of Public Health, 3*(3), 392–396.

Flicker, S., Savan, B., McGrath, M., Kolenda, B., & Mildenberger, M. (2007). "If you could change one thing . . . ": What community-based researchers wish they would have done differently. *Community Development Journal, 43*(2), 239–253.

Kindig, D., & Stoddart, G. (2003). What is population health? *American Journal of Public Health, 93*(3), 380–383.

Ibáñez-Carrasco, F., & Riaño-Alcalá, P. (2011). Organizing community-based research knowledge between universities and communities: Lessons learned. *Community Development Journal, 46*(1), 72–88.

Labonte, R., Muhajarine, N., Abonyi, S., Woodward, G. B., Jeffery, B., Maslany, G., et al. (2002). An integrated exploration into the social and environmental determinants of health: The Saskatchewan Population Health and Evaluation Research Unit (SPHERU). *Chronic Diseases in Canada, 23*(2), 71–76.

McIntosh, T., Jeffery, B., & Muhajarine, N. (2010). Introduction: Moving forward on critical population health research. In T. McIntosh, B. Jeffery, & N. Muhajarine (Eds.), *Redistributing health: New directions in population health research in Canada* (pp. xi–xxv). Regina: CPRC Press.

Roche, B. (2008). *New directions in community-based research*. Toronto: Wellesley Institute.

Rogge, M. E., & Rocha, C. J. (2005). University-community partnership centers. *Journal of Community Practice, 12*(3), 103–121.

Ross, L. F., Loup, A., Nelson, R. M., Botkin, J. R., Kost, R., Smith, G. R., & Gehlert, S. (2010). The challenges of collaboration for academic and community partners in a research partnership: Points to consider. *Journal of Empirical Research on Human Research Ethics*, 19–31.

Salmon, A. (2007). Walking the talk: How participatory interview methods can democratize research. *Qualitative Health Research, 7*(7), 982–993.

Sandmann, L.R. (2008). Conceptualization of the scholarship of engagement in higher education: A strategic review, 1996–2006. *Journal of Higher Education and Engagement, 12*(1), 91–104.

Savan, B. (2004). Community-university partnerships: Linking research and action for sustainable community development. *Community Development Journal, 39*(4), 372–384.

Savan, B., & Sider, D. (2003). Contrasting approaches to community-based research and a case study of community sustainability in Toronto, Canada. *Local Environment, 8*(3), 303–316.

Stanton, T. (2008). New times demand new scholarship: Opportunities and challenges for civic engagement at research universities. *Education, Citizenship and Social Justice, 3*(1), 19–42.

Tyron, E., Stoecker, R., Martin, A., Seblonka, K., Hilgendorf, A., & Nellis, M. (2008). The challenge of short-term service-learning. *Michigan Journal of Community Service Learning, 14*(2), 16–26.

University of Regina. (2011). *Working together towards common goals: Serving through research*. Strategic Research Plan 2010–2015. Regina: University of Regina.

University of Saskatchewan (2006). *Foundational document on outreach and engagement: Linking with communities for discovery and learning*. Saskatoon: University of Saskatchewan.

University of Saskatchewan (2008). *The second integrated plan: Toward an engaged university, 2008–12*. Saskatoon: University of Saskatchewan.

University of Saskatchewan (2011). *Third integrated plan: Areas of focus*. Saskatoon: University of Saskatchewan.

University of Saskatchewan (2012a). *Promise and potential: The third integrated plan 2012–2016*. Saskatoon: University of Saskatchewan.

University of Saskatchewan (2012b). *Engaging with external partners. Recommended principles, guidelines, and action plan components*. Prepared by the Working Group on Engaging with External Partners. Saskatoon: University of Saskatchewan.

Wallerstein, N. B., & Duran, B. (2006). Using community-based participatory research to address health disparities. *Health Promotion Practice, 7*(3), 312–323.

1
ETHICS OF COMMUNITY-BASED RESEARCH

WORKING TOGETHER
ETHICAL PRACTICE IN COMMUNITY-ENGAGED RESEARCH

Diane Martz and Juanita Bacsu

Introduction

Community-engaged research describes research partnerships between academic researchers and communities that can take a variety of forms, ranging from an equal partnership in all phases of research to partnerships controlled by community or academic researchers (Ross et al., 2010). This chapter focuses on the lessons learned about ethical research practice by members of the Saskatchewan Population Health and Evaluation Research Unit (SPHERU) and the Community-University Institute for Social Research (CUISR) and their community research partners over the past decade of community-engaged research. The term "community-engaged research" is appropriate to describe the research projects reported in this chapter since they range from community-based participatory research (CBPR), in which the community and university partners were joint decision makers, to projects in which the community was the major decision-making partner and other projects in which the academic researchers were the main decision makers.

Over the past ten years, discussions on the particular ethical issues and considerations inherent in community-engaged research have begun to bear fruit. The principles that have informed research ethics guidelines around the world were first laid out in the 1979 Belmont Report and incorporated into the first (1998) *Tri-Council Policy Statement on Ethics in Human Research: Ethical Conduct for Research Involving Humans* (TCPS). These initial Tri-Council guidelines were soundly criticized for their biomedical approach to research, which not only ignored the more qualitative focus of much community-engaged research but in many cases made it difficult to carry out community-engaged research (Flicker, Travers, Guta, McDonald, & Meager, 2007). The underlying ethical principles in the TCPS document were based on notions of human dignity, justice, respect for free and informed consent, vulnerable persons, privacy and confidentiality, and balancing harms and benefits. Although these are worthwhile goals, they focused on the individual as a research subject and did not anticipate questions of social justice, community harm, or community consent (Flicker et al., 2007; Reid & Brief, 2009). Furthermore, these guidelines did not anticipate that risks from research participation can be different at the levels of the individual and the community. Several authors have noted that the traditional focus on individual ethics has given rise to a number of ethical research issues that might harm marginalized communities, including community harms, inappropriate research methods, communities being over researched, coercion, findings not being shared with communities, and communities being stigmatized through their participation in research (Flicker et al., 2007; Israel, Schulz, Parker, & Becker, 1998; Ross et al., 2010; Schnarch, 2004). The second edition of the *Tri-Council Policy Statement on Ethics in Human Research: Ethical Conduct for Research Involving Humans* (TCPS2), released in December 2010, was the product of considerable consultation in Canada. These guidelines recognize to some extent that research takes place with individuals as well as groups and communities, and highlights the importance of respect for persons, concern for welfare, and justice.

Although community-engaged research with strong community and academic partnerships has had some success in addressing the above ethical issues, Flicker and colleagues (2007) point to new ethical questions, which include determining who is the appropriate representative of a community, conflict between community members, power relationships, consent at the community level, confidentiality and access to sensitive data in a partnership, data ownership and control, and balancing processes and outcomes. This chapter examines the experiences of academic and community researchers from SPHERU and CUISR in the context of widely accepted ethical principles of respect for persons, concern for welfare, and justice. We are particularly interested in the extent to which these principles are considered as ethical issues and dilemmas as they arise during research projects.

The information for this chapter was gathered through qualitative interviews with six academic and four community researchers associated with CUISR and SPHERU. The experiences of the researchers ranged from projects within a single geographical community to projects involving many different geographical communities. They have worked in urban, rural, and Aboriginal communities. In some cases, the projects were initiated by the community, and academic researchers were brought in later; in other cases, academic researchers initiated the relationship.

TCPS2 (CIHR et al., 2010) is based on the principles of respect for persons, concern for welfare, and justice. Respect for persons, based on the right to self-determination, is ensured through free, informed, and ongoing consent of the participant to be involved in the research. Respect also speaks to the need to ensure protections for the community and for those who lack the capacity to grant consent for themselves, and it recognizes that power relations can influence consent and participation in the research. The second principle, concern for welfare, is about maintaining the most favourable balance of risks and benefits, fully informing participants about risks, and not exposing them or their communities to unnecessary risks. Risks can be physical, economic, social, or psychological and involve considerations of privacy and the control of information. The third principle, justice, refers to the need to treat all people fairly and equitably in terms of participation in and benefit from research. Furthermore, those who live in vulnerable or marginalized circumstances might need special attention to be treated justly in research. Within this principle is the recognition that there might be a power imbalance between researchers and participants.

Respect and Self-Determination

When research is focused primarily on individuals, self-determination refers to the requirement to be fully informed and able to consent freely to participate in research. In community-engaged research, in which the community can be both research partner and participant, self-determination is found through meaningful participation in all aspects of the research process. The community and academic partners interviewed saw building and sustaining respectful relationships as a key element of success in community-engaged research. Respect was mentioned a number of times as a critical component of building trust, and researchers talked about developing respect by creating an environment respectful of the various partners. For example, within the research partnership, there can be competing interests among the community, the funders (government departments, granting agencies), and the researchers. This was especially noted in evaluation research projects, in which the community often has an interest in keeping a program, the funder has an interest in determining if the program is meeting its objectives, and the researcher has an interest in doing valid research.

Other aspects of building trust included showing respect for the cultural issues that people bring to the table and demonstrating a shared commitment to change. In many of the projects, trust in an academic researcher was facilitated by the researcher being sponsored to enter the community by a trusted and respected community member. Trust was also thought to be influenced by the reputation of the researcher and the institutions with which researchers are affiliated.

An important element of building trust is having a presence in the community. For example, one academic researcher commented that researchers have learned from experience that frequent shorter visits are better, while another noted the need for constant interaction with the community and other members of the research team. Academic researchers also noted the following.

- *By going out to communities, you have to be out there, they have to get to know you.*
- *You can't build trust if you're not visible, if they don't know who you are, if they don't know what you are about, if they don't believe you are committed to their community.*

Both community and academic researchers noted the need to be inclusive in the research project and ensure that everyone believes his or her contribution is important. Moreover, it is important to recognize that both academic and community researchers bring a particular set of skills and knowledge to the table. As one academic researcher commented, "I rely on the knowledge of communities; it [the research] has to be relevant, something they can use, that is why they have to be involved from the start." Some researchers would go further, believing that true collaboration results when the community takes leadership of a project.

The researchers interviewed reported that some of their early concerns about working together were laid to rest as the relationship evolved. One academic researcher anticipated being told "things that I would expect to be told, not really included, being peripheral to their activity, and given entrance as the community saw fit." Another academic researcher reported that, "when I first started, I thought I had to be the expert, and I had to know everything, but I have become more confident, more confident to say to them 'I don't know' or 'I don't understand' or 'Why are we doing that?'"

A community researcher was initially concerned that colleagues might feel "second best" because "they would be with all these masters and doctors." They were surprised that the academics "had a wonderful sense of humour, were very warm and welcoming, and made [the community researcher] feel happy that you were there and that they were happy that you were there."

Researchers also talked about their challenges negotiating differences within the community-academic research team, such as differences in terminology and language. For example, researchers noted difficulties in understanding when partners used academic language or disciplinary specific terminology, and when the project involved research team members and participants with a variety of first languages. Moreover, they experienced differences in familiarity with and knowledge of the research process, leading to questions about data analysis and different expectations and assumptions about how the group would work together and make decisions, as well as what were desirable outcomes of the research process.

The key to resolving these challenges was open and transparent communication. Researchers commented that, for the relationship to work, there need to be conversations about assumptions and expectations, how the researchers will work together, how they will communicate. One academic researcher strongly recommended negotiating a written agreement or memorandum of understanding between the researchers and the community. The content of an agreement will vary from community to community but can include guidelines for the communication of results (who writes them, who releases them, who decides when they are ready to become public, which format they should take), how researchers will keep in touch with community people, the governance structure, and the roles of each partner. Ross et al. (2010) suggest that having these detailed discussions and agreeing on a memorandum of understanding are akin to a free and informed process of consent for the community.

One of the most common issues that arose was time. Most of the researchers interviewed commented that it takes considerable time to gain an understanding of the community and to build substantive and trusting relationships with community research partners and the community at large. "It's worth every minute. If you really truly want to be involved and you open yourself up to it,...it is the most honest and enriching form (of research),... but do be prepared for that long-term investment." Academic researchers work on relatively long time frames, and every phase of a research project— from grant writing, response to a grant proposal, research design, data collection, analysis, to writing—can take months to complete and often seems very slow to community research partners. Community organizations might be working with a different set of timelines involving annual reviews, deliverables, contracts, and program funding, and often they are adding their work on a research project to their already busy schedules. One community member mentioned feeling guilty that she could not commit 100% to the project because "issues would arise at work" that resulted in missed meetings. Researchers commented that "it is a balancing act to engage people, but not overload them," and that "community people's time must be respected." Another mentioned the importance of making the research worthwhile for community partners by making sure that the findings get back to the people and that commitments made to the community are kept. Time is also important in the sense of a long-term commitment to a relationship with the community and in the resources needed to accomplish it.

Concern for Welfare

Concern for welfare speaks to a favourable balance of risks and benefits for individuals and communities. Both community and academic researchers saw significant benefits to being part of a community-engaged research project. The community researchers thought that engaging with academic researchers gave them increased access to resources, such as extended networks, grant money, and expertise. One community researcher found opportunities to connect with similar projects in other communities through contacts established by the academic researchers. The long-term relationship building and networking had benefits for subsequent projects and an opportunity to be invited into the university classroom to share experiences from the communities.

In many of the projects, the community researchers reported that they came to appreciate the value of research and what it has to offer for communities and their organizations. Communities built networks with other groups and relationships with researchers. Community researchers reported that they learned how to use community-engaged research as an advocacy tool and learned a lot about the system in which they were operating. At the individual level, one academic researcher reported that the project built confidence in the participants and pride in the project and its outcomes. Some projects built research skills among community members, although an academic researcher noted that the high level of turnover of research assistants meant that the skills did not always stay in the community. From the perspective of academic researchers, being part of community-engaged research was "a very privileged position" and an opportunity to "make a difference." It was an opportunity for this researcher to be "part of something rather than a voyeur or onlooker."

Academic researchers also thought that partnering with the community gave them more credibility and better access to community members. Developing a long-term relationship with the community allowed for a deeper understanding of the issues, challenges, and solutions put forward by the community. "Authentic research material is truly built on those relationships that you've taken time to foster and grow." Academic researchers also talked about doing research that is relevant to the community, in particular being able to work on issues of concern today, to frame research questions in ways that make sense to communities, to collect information in ways that are acceptable and workable, and to give information back to the community in ways that are useful.

Risks to the Community

Ross et al. (2010) explore in detail the ethical considerations of community-engaged research and propose a taxonomy of risks that summarizes potential risks to the individual, the members of a group participating in research, and the community itself. Risks to well-being can arise for each individual or group both in the process of doing research and in the outcomes of the research. Risks to the individual in the process of research include the risk of either self-identification or external identification as a member of a particular group. Risks to communities from the process of research include disruption of internal relationships. Risks can also arise from the outcomes of research when negative findings create disruptions in the community or damage the reputations of individuals, groups, and communities.

Risks to agency can arise for individuals, groups, and communities from the outcomes of community-engaged research. Risks to individual agency are typically managed through the process of consent and formalized in signing the consent form. Risks to group or community agency can occur when the leadership or moral authority of a group or community is undermined by the actions of academic researchers. These risks can be addressed by the relationship building inherent in forming a research partnership and can be formalized by the development and signing of a research agreement or memorandum of understanding (Khanlou & Peter, 2005; Ross et al., 2010).

Researchers noted that there were risks to their community partners in the research process, specifically risks to the relationships involved if one group had negative things to say about another group involved in the research. They also noted risks to the community if the research outcomes showed problems rather than solutions or deficits rather than strengths. Some of the projects involved funding risks for the program being evaluated if the research results indicated the program to be ineffective.

Research teams dealt with these risks by building trust early in the project and setting in place the kind of relationship in which they could anticipate issues and "really talk those issues through." Community researchers mentioned that they dealt with potentially negative research outcomes by not constructing negative research questions and planning a communication strategy that would portray research results as proactive rather than reactive.

Researchers also talked about the individual risks that can arise, such as collecting data that can identify people, triggering sensitive memories, and creating safe places for people to talk about issues. Researchers noted that they worked hard to ensure that confidentiality was maintained and that people had voice. One community researcher commented that it was not always possible to ensure that people had voice when service providers and government representatives were present at community meetings; some were silenced, although others were not. However, they also noted that not all people want to be protected, that some citizens think they can take care of themselves and will confront powerful institutions.

Risks to Researchers

Ross et al. (2010) also argue that researchers can be exposed to risks through community-engaged research and discuss the threats to individual moral agency that arise from risks to reputation and trust. These reputational risks extend to their institutions and the broader research community.

Researchers recognized that erosion of trust can result if one's actions undermine the relationship or a researcher introduces someone to the project who does not put the same importance on ethical work and behaves unethically. "Doing the work ethically means slowing down enough that you can deal with the kind of ethical challenges that arise." One academic researcher stated that his colleagues were becoming involved in fewer large projects because of the difficulty of "maintaining good communications" and noted that "trust erodes very quickly." This researcher looked for a research relationship in which there was both personal and professional trust among academic researchers.

Many community-engaged research projects involve high expectations from the community partners. As one community researcher told us, one project got people in the community "thinking and dreaming a bit, identifying new ideas," and it gave them "a bit of hope." Roche (2008) points out that both community and academic researchers must have realistic assessments of the research outcomes because there are reputational risks if expectations are too high or if something useful is not delivered on time.

You have to do what you say you are going to do because people are going to judge you on your ability to follow through on what you said you would do. Researchers making promises they haven't kept is at the root of many bad experiences in communities, especially First Nations communities and even rural communities.

In this research, community and academic researchers identified additional risks regarding their employment. One community researcher noted that insisting on a more community-engaged approach in a research project created the possibility of losing her job. She also felt caught in the crossfire between partners and funders of the project. Many authors have commented that universities commonly value community-engaged research less than traditional research (Roche, 2008). An academic researcher noted that community-engaged research does not fit well with the system of tenure and promotion at universities. Community-engaged research takes more time, and publications must be created in various forms that are useful to the community and not just academic journals. As a result, a researcher might be assessed as having lower research productivity, negatively affecting opportunities for promotion.

Other risks identified included psychological risks to community researchers as they struggled with uncomfortable stories and situations and how to respond to them. Psychological risks were handled by debriefing both formally and informally through intensive team discussions. Researchers also noted personal risks in some interview situations or from being known to be involved in a controversial project.

Justice
In the context of research ethics, the principle of justice refers to ensuring equal access to the opportunity to participate in research and to benefit from the outcomes of research. In community-engaged research, with its orientation toward social change, the idea of justice is closely related to concepts of social justice (Shore, 2006). Both perspectives were evident in the experiences of the community and academic researchers. From an equity perspective, community researchers talked about how important it was to be inclusive in every possible way and to make sure that barriers to participation were removed, whether they were due to financial, language, or safety considerations. One academic researcher asked, "When you choose a particular gatekeeper, how do you ensure everyone in the community has an equal chance of participating?"

Considerations of social justice were evident in questions posed by community researchers. "Is the research a worthwhile way to spend public money?" "Is it a question that deserves special treatment?" In one project, the research team explicitly considered social justice when they "took on an issue we thought was very important in terms of justice for a group of people." Social justice is also expressed in terms of the benefits cited by community researchers, such as learning how to use community-engaged research as an advocacy tool, and by academic researchers in their desire to conduct research that can make a difference.

Researchers indicated that they dealt with ethical issues that arose throughout the projects by trying to anticipate them and by talking through them when they arose. The interviews indicated that community and academic researchers used a principled approach that incorporated the basic tenets of respect, welfare, and justice. One researcher noted that the team worried about many possible ethical issues that "did not materialize."

To manage ethical issues, researchers drew on resources such as the experiences, skills, emotional intelligence, and principles of their research teams and advisory committees. They also benefited from mentorship from other organizations, academics, and Elders involved in the project. One academic researcher noted that the CIHR *Guidelines for Health Research Involving Aboriginal Peoples* (CIHR, 2007) was a good resource. Another commented on experiences conducting research with Aboriginal communities, where discussion on engaging in a research project "always begins with a conversation about ethical issues of engagement [as] part of the get-to-know-you." Another researcher thought that her team would have benefited from connecting with the university Research Ethics Board for help with some of the complex issues regarding personal disclosure in focus groups and public meetings. Guillemin, Gillam, Rosenthal, and Bolitho (2010) suggest that institutional ethics committees are an untapped resource and could be much better utilized by researchers.

Conclusion

Community-engaged research is a complex journey, often requiring continual negotiation and dialogue among multiple stakeholders and partners. Effective community-engaged research incorporates ethical guidelines to help protect the needs and interests of participants, researchers, and communities. The key ethical principles of respect, concern for welfare, and justice contribute to ensuring a balance among the perspectives, values, and interests of community and academic researchers, participants, and communities. These key ethical concepts are not just important when thinking about the requirements for ethics review by the Research Ethics Board or Institutional Review Board but also guide the decisions and actions of a group of researchers in their research practice. When asked about the three principles of respect, concern for welfare, and justice, six of the ten researchers interviewed talked specifically about them in the context of their research projects. As well, the principles were evident in the researchers' discussions on building trust, establishing processes of consent, dealing with concerns about confidentiality, ensuring equal opportunities to participate in research projects, and addressing the diverse needs of various stakeholders. Most often these concerns were addressed through open and transparent communication, including negotiating a written agreement to address ethical issues, terms of the research process, and roles and responsibilities of the different partners.

Considering ethical principles helps researchers to establish and facilitate shared power, equal voice, and collective decision making among partners. Recognizing the importance of ethical concerns throughout the research journey aids in building trust and sustaining the research partnership. Moreover, recognizing that ethical principles extend beyond the individual to the community will help to support successful research outcomes.

References
Canadian Institutes of Health Research (CIHR). (2007). *CIHR guidelines for health research involving* Aboriginal peoples. http://www.cihr-irsc.gc.ca/.
Canadian Institutes of Health Research, Natural Science and Engineering Council of Canada and Social Sciences and Humanities Research Council of Canada. (2010). *Tri-Council policy statement: Ethical conduct for research involving humans.* http://www.pre.ethics.gc.ca/.
Flicker, S., Travers, R., Guta, R., McDonald, S., &'Meager, A. (2007). Ethical dilemmas in community-based participatory research: Recommendations for institutional review boards. *Journal of Urban Health: Bulletin of the New York Academy of Medicine, 84*(4), 478–493.
Guillemin, M., Gillam, L., Rosenthal, D., & Bolitho, A. (2010). Resources employed by health researchers to ensure ethical research practice. *Journal of Empirical Research on Human Research Ethics, 5*(2), 21–34.
Israel, B., Schulz, A., Parker, E., & Becker, A. (1998). A review of community based research: Assessing partnership approaches to improve public health. *Annual Review of Public Health, 19*(1), 173–194.
Khanlou, N., & Peter, E. (2005). Participatory action research: Considerations for ethical review. *Social Science and Medicine, 60*, 2333–2340.
Reid, C., & Brief, E. (2009). Confronting condescending ethics: How community-based research challenges traditional approaches to consent, confidentiality, and capacity. *Journal of Academic Ethics, 7*, 75–85.
Roche, B. (2008). *New directions in community-based research.* Toronto: Wellesley Institute.
Ross, L. F., Loup, A., Nelson, R. M., Botkin, J. R., Kost, R., Smith, G. R., & Gehlert, S. (2010). The challenges of collaboration for academic and community partners in a research partnership: Points to consider. *Journal of Empirical Research on Human Research Ethics, 5*(1), 19–31.
Schnarch, B. (2004). *Ownership, control, access, and possession (OCAP) or self-determination applied to research: A critical analysis of contemporary First Nations research and some options for First Nations communities.* Ottawa: First Nations Centre, National Aboriginal Health Organization.
Shore, N. (2006). Re-conceptualizing the Belmont Report: A community-based participatory research perspective. *Journal of Community Practice, 14*(4), 5–25.

TALKING TO THE "HEALING JOURNEY" INTERVIEWERS
ETHICAL CONCERNS AND DILEMMAS

Wendee Kubik, Mary Hampton, Darlene Juschka, Carrie Bourassa, and Bonnie Jeffery

Introduction

In 2004, a number of university researchers and their community partners from the three prairie provinces received a $1 million Social Sciences and Humanities Research Council of Canada (SSHRC) grant through the Community-University Research Alliance (CURA) program to conduct a five-year longitudinal study of women who had been abused by their intimate partners. This study, entitled "The Healing Journey: A Longitudinal Study of Women Who Have Been Abused by Intimate Partners," was led by Dr. Jane Ursel from the University of Manitoba through RESOLVE (Research and Education for Solutions to Violence and Abuse).

As a CURA, the study was designed and implemented by academic researchers and community partners. For example, in Saskatchewan, six academic researchers from Regina, Saskatoon, and Prince Albert were teamed with community partners who delivered anti-violence programs through Saskatchewan Justice as well as transition houses. The structure of the CURA set up a team in which each province had an academic and a community coordinator at each site. When the recruitment strategy for the "Healing Journey" was designed, we first conducted focus groups with experiential women (i.e., women who had experienced intimate partner violence) as well as front-line service providers to design a recruitment strategy that would be respectful of and successful with this hard-to-reach, vulnerable population. Based on their recommendations, a recruitment strategy was adopted whereby community partners would refer eligible and interested women to academic partners, who would then contact the women and go through the consent procedures with them. Although each province handled its own recruiting and interviewing, overall management of the study and the tri-provincial database was handled by the principal investigator at the University of Manitoba. Monthly teleconferences helped each province to stay on track with the overall interviewing schedule. Each province (Alberta, Saskatchewan, and Manitoba) set a goal to interview 200 women every six months with the intention to understand the factors involved in women's survival of and healing from partner violence and abuse. The aim was to inform service providers and policy-makers about effective programming and gaps in services for women involved in abusive relationships and to broaden knowledge in the field.

The "Healing Journey" project presented several challenges. As investigators and project directors, we were responsible for the safety of the participants (many of whom had just left an abusive, dangerous situation) as well as the interviewers, who were interviewing a vulnerable population. We needed to take precautions so that participants were not endangered by participation in the study, such as their whereabouts being revealed if they were in hiding. We also needed to ensure that none of the interviewers was placed in a dangerous situation, and we were frequently called on to help them cope with the often distressing and at times disturbing stories related by the women. The subject matter of the interviews (intimate partner abuse faced by the women) is a personal and very sensitive area, and there was a need to approach the interviews with empathy so as not to revictimize the participants. Many of the women felt stigmatized from their situations and were hesitant to talk about aspects of their abuse until trust was established with the interviewers. Because of these challenges, we faced a number of ethical concerns in conducting this kind of research.

As the research progressed, we became aware that the interviewers were having a unique research experience and had begun to amass a great deal of knowledge about conducting research with hard-to-reach, vulnerable populations. The interviewers themselves realized that they had this "storehouse of knowledge" of unexpected and valuable insights regarding research with at-risk populations and wanted to share it with us.

To learn about the experiences, concerns, and ethical dilemmas of the interviewers, a smaller project called "Interviewing the Interviewers" was developed by five academic researchers at the University of Regina and First Nations University who were involved in the "Healing Journey" project in Saskatchewan. Thirteen semi-structured, in-depth interviews were conducted in Regina, Saskatoon, and Prince Albert with the Saskatchewan interviewers in the summer of 2009. The interviewers were asked a series of questions about their experiences, reactions, relationships with the participants, insights gained, and problems encountered.

The goal of this chapter is to highlight some of the areas discussed during these interviews, including personal boundaries, methods of self-care, barriers to interviewing, issues of safety and confidentiality, cultural sensitivities, knowledge gained, and training of and support for interviewers.

Background/Context

A key aspect of the "Healing Journey" was the involvement of community partners in all phases of the study. A number of community partners in each province (e.g., women's shelters, provincial agencies providing services to abused women, transition houses) helped to recruit participants, gave invaluable advice about recruitment and retention, and provided strategies for interviewing and keeping in contact with the participants. Several community partners allowed interviewing in their facilities and provided counselling and debriefing for the participants if needed. This was an integral part of the study and made the project possible because of community partners' knowledge and support. These partners were regularly consulted and became involved in our discussions and academic meetings, and they complemented the academic research expertise with their in-depth practical knowledge. We constantly sought their advice on the best way to go about our research. In turn, we shared the information from our study with community partners. Some questions on the interview schedules came directly from these partners since these were areas of interest to their work. Without the involvement of community partners, we would not have been able to complete the study.

Philosophical assumptions behind the "Healing Journey" played a key role in how the research was conceptualized and how the interviews were conducted. Interviewers were hired by academic and community researchers; qualifications for interviewing included some experience with or sensitivity to feminist issues and the context of intimate partner violence. Most of the interviewers were university-trained undergraduate and graduate students; a few were community-based shelter workers. Others had experience working in women's shelters or with women who had been in abusive situations. We relied heavily on the interviewers to maintain a caring and ethical research relationship with the women in the study, so they were chosen carefully. Interviewers contacted participants every six months for a face-to-face interview of two to three hours. Throughout this process, various ethical challenges emerged. For example, since the participants were victims of intimate partner violence, confidentiality was particularly important. Some of the women were in hiding, so the interviewers had to develop innovative methods to maintain contact. In other instances, staying in contact was always a challenge since some of the participants lived in northern and remote communities, while others, due to the realities of intimate partner violence, moved frequently.

Following are some of the main issues that the interviewers encountered while working on the "Healing Journey." Quotations from the interviewers themselves are used to illustrate the problems that they faced.

Retention and Safety Issues

Two of the biggest challenges faced by the interviewers were retention of the women and safety issues. Because many of the women were often in hiding, issues of safety and confidentiality were particularly important, so the interviewers developed methods for maintaining contact and ensuring the safety of the participants along with their own safety. One method was safe networking. It consisted of providing the interviewer with a list of safe contacts through which to reach the participant if she moved with some frequency or went into hiding. The community partners were also very helpful because they would pass along messages to the participants if they came through the services again. The interviewer would block the number that she was calling from so that the participant's phone would not display the number. Several participants did phone interviews since they felt safe in their homes and it was easier for them to take part this way. At times, there were delays between the different waves of interviewing, and the interviewers found that letting the women know about the delays and affirming they still needed them in the study helped to keep the women involved. The "Healing Journey" also had a main telephone number given to all the women so that they could contact us at any time. Participants appreciated updates on the study since they were motivated to see the results. Interviewers had mobile phones with them and let the project coordinators know the time and place of each interview.

Communication and flexibility were then seen as key to keeping women in the study; some women wanted interviews to be set up by e-mail, others by phone, others through the agencies, so finding the right and safest way to connect with the participants led to good retention. Participants were given an honorarium of fifty dollars per interview, not as an incentive since this was such a little amount, but in recognition of the contributions that they were making. University policy required a certain protocol for disseminating honoraria that did not honour anonymity, so we negotiated with financial services to make available cash honoraria for the participants with anonymous signatures when they received the funds.

Boundaries and Ethics

Although the interviewers were able to set and maintain boundaries, many struggled between being "counsellor" versus "interviewer." In addition, many felt close bonds with the participants, and though they maintained boundaries it was difficult in many instances to do so. One of the interviewers commented that

> we were told as interviewers we had to remain neutral. I found that sometimes cold—you know normally if you are a person and you are talking to somebody and they are having a hard time you can pat their shoulder and you can say something to them to kind of reassure them, but we were cautioned to not get too close or get too wrapped up to allow ourselves to do that. So, for myself, I would just offer them a break, or I would say, "Take your time, don't worry about it, it's ok."

Maintaining boundaries was an ongoing issue and was discussed regularly, and the project coordinator was available to talk to the interviewers about the issue.

Experiences of the Interviewers

All of the interviewers had experiences that they said will stay with them. Some were humorous, joyful, or hopeful; others were threatening and at times traumatic. Some of the experiences that the interviewers mentioned are as follows.

Grief

> [She was] unstable, severely depressed, and sitting at her kitchen table . . . sobbing about how she feels like killing herself; you can't remain neutral in that moment, like you absolutely can't.

> I didn't have any counselling experience yet at that point either. . . . I would say definitely that would have been an example of some vicarious trauma. When I got home from that interview, I just burst into tears because the feelings that I had were mostly of guilt.

Hope

> The desire of the women to really see change, that is so heartening, that is something that I will continue to look back on and admire.

> I just wanted to weep for her. It's just like why wouldn't you give up, . . . of course you want to give up, of course you just want to throw in the towel? That's one thing I still think about. . . . But the thing that was good about it was that the next time I talked to her there had been some movement, and she was hopeful, and even by the end of that interview I felt like I did help her in some way.

Humour

> I asked, "How is your life as a whole?" I just read the question, and there was no answer, and so I looked up, and I waited, and she was just kind of glaring at me, kind of upset with me. All of a sudden she started laughing, and she said to me, "Oh, my goodness, you know what I thought you said, I thought you said, 'How is your life as a whore?'"

Managing Personal Responses

Many of the interviewers heard horrific stories of violence. Although expecting to hear such stories and being trained to cope with these experiences, they still often found them traumatic and had to cope emotionally. When training the interviewers, particularly for the qualitative interviews when details of violence would be narrated, significant workshop time was devoted to learning to establish personal boundaries between interviewer and participant. These boundaries were intended to protect both the interviewer and the participant insofar as the interviewer risked being traumatized by the narrative of the participant and the participant risked being betrayed by the interviewer's response to her story. The workshop advisers emphasized the need for vigilance in monitoring one's own well-being, recognizing if and when a personal boundary had been breached. To do so, the interviewers were advised to talk to each other about their experiences, to meet with the interview coordinator, to come together as a group with the interview coordinator to discuss interviews, and to meet with any of the researchers who had a background in counselling and/or therapy. With these mechanisms in place, the interviewers took advantage of coming together to "debrief" and to meet with the interview coordinator. From these "debriefing" sessions and the sharing of knowledge (while adhering to the ethical guidelines for confidentiality), the interviewers were supported and developed coping mechanisms. Following are some statements that the interviewers shared regarding how they coped.

I think you have to be a strong person because the stories you hear are difficult. I would think that I had heard probably the worst story that I could hear, and then I would do another interview, and I would hear something worse. You know? It just made me realize how resilient these women are and how strong because sometimes, and of course I couldn't say anything, but I was thinking to myself, I don't even know how you can be sitting there talking to me when you have been through all of this and you know I'm wondering how I would be, could I do this? Could I be you?

A lot of times, if I am really affected by an interview, and I learned to do this in my other work when I feel really jammed up or have picked up too much energy, I structure this. I will go home, I will light candles, I will fill up the tub, and I really worked myself into having a good cry, just released some of that energy, and in a prayerful sort of way just released that back to the universe, that it is too big for me to hold, pray for her. So that's how I've worked with it.

Because the interviewers wanted to help but were not always aware of the resources available in different areas of the province, a scan of resources for women experiencing violence or related problems was developed. Interviewers could then refer women to local resources if they were available. Although the guide was helpful for the interviewers and participants, it was frustrating when there were no resources available in a particular area.

Cultural Sensitivities

Some of the women interviewed were of Aboriginal ancestry. Although a number of the interviewers were also of Aboriginal ancestry, there was a need for understanding the diversity of Aboriginal people and culture. When we trained the interviewers, we discussed cultural sensitivities and talked about the cultural diversity of Aboriginal people and the historical effects of colonization on the contemporary realities of Aboriginal people, in particular Aboriginal women. Despite this training, some interviewers struggled to establish a relationship of trust, and that is only natural. One cannot expect hundreds of years of damaged relationships to be repaired overnight. We encouraged the interviewers to be patient and reminded them that some of these women had suffered multiple oppressions and intergenerational trauma. We told them that they could not force a relationship, nor should they take the lack of a relationship personally. Interviewers needed to be aware of the reality of hundreds of years of colonialism and understand what was in the realm of possibility for this project.

Developing a trusting relationship required that the interviewers be flexible, patient, willing to hear the participants' stories, non-judgmental and open to the participants, and, most importantly, willing to let the participants take the lead in telling their stories. Interviewers were often "matched up" in terms of age, or Aboriginal interviewers would interview Aboriginal women, but understanding the effects of colonialism was important in giving context to the interview.

> *The woman [whom I interviewed] dealt with residential schools and things like that. I felt it probably would have been better if she had an Aboriginal interviewer because she would go on a tirade, justly, about these white people, and I felt bad for her to have to. Maybe myself as a white woman interviewing her may have felt manipulative or coercive, or in that case I thought I recognize that as a white*

person we have so much, I have so much, privilege that in her case perhaps my being white, which I often take for granted, was a detriment to her. That was a really good experience for me, but I'm not sure for her.

Flexibility is the key. You know, being open to understanding the way they do things with regard to tobacco giving. It's not necessarily a Cree tradition, I mean it's not just tobacco, it could be cookies, or in some cases you ask the person what they would like, and it could be a chicken dinner, you know what I mean? We are so hung up on these types of things, right, but it is basically dependent upon the woman.

The "Healing Journey" had an Elder as a team member who gave important advice regarding training of the interviewers and the project itself as it evolved. Another of the researchers was Métis, and the project coordinator was able to consult with her about several of the issues that the interviewers encountered.

Confidentiality

Confidentiality was an ethical concern because we did not want to further endanger the women, particularly if they were in abusive relationships. Trust between interviewer and participant became a paramount concern throughout. As feminist researchers, we recognized that we did not want to "exploit" the women; however, at times, we did need to ask probing questions. Confidentiality and anonymity were always issues, and researchers and interviewers kept them in mind throughout the project. Confidentiality and anonymity were maintained through standard methods (as stipulated by the Research Ethics Board) such as providing numbers for participants' names in all databases, and front-line service providers who had referred participants to the study never knew for sure which of their clients had decided to participate in the study unless the participants told them. In all debriefing sessions with the interviewers, confidentiality was constantly emphasized.

Outcomes/Lessons Learned

Community and academic members worked together to develop all aspects of this longitudinal project, including thinking about the interviewers and the interview process itself. Members of the project recognized from the outset that the process of interviewing would be central to the success of the project; furthermore, we knew that the study would become a part of our participants' journeys. In light of these two recognitions, then, it was very important to think about what our intersections with the participants' lives ought to look like.

We focused on a number of areas that we thought might contribute to rather than detract from the lives of the participants: the interviewer and her ability to situate herself in the study and the groups involved; her ability to represent the study to the participants; how to compassionately and respectfully deliver questionnaires to the participants and listen to their stories; how to recognize and deal with distress of the participants as they told their stories; and how to extend to these women our appreciation for what they were sharing with us. These areas required reflection on our part. Since students were the primary researchers, we worked hard to ensure that we had a good mix of younger and older female students and that they came from a variety of cultural backgrounds, including Métis and other Aboriginal populations since we anticipated the inclusion of Métis and other Aboriginal women. Having determined the interviewers, to prepare them, researchers and community partners developed a workshop to discuss and practise the interviewing process.

During the morning segment of the workshop, discussions examined the challenges of interviewing and envisioned the kinds of issues that could come up, such as miscommunication, a traumatic response to a question, and safety. Also included in this first part of the workshop were discussions on the mixed cultural, economic, educational, and class backgrounds of the anticipated participants and the challenges that they faced in their lives. As feminists, we were highly conscious of power differences that function in societies, and this aspect of our workshop comprised an effort to increase an interviewer's understanding of these differences.

The afternoon component of the workshop was comprised of role-playing and acting out interviews. Initially, several interviews were acted out by the facilitators of the workshop, and the participants were asked to identify what they perceived as viable and non-viable with regard to the interview process. Thereafter, in groups of two, each group devised and acted out an interview while others watched and responded with insights and questions. Finally, discussion closed the workshop, and from it an interviewer protocol and guide were developed.

Establishing a workshop, protocol, and guide for the interviewers, although not original, is often a neglected aspect of people-centred studies. In the case of the "Healing Journey," however, this aspect occupied our attention and reflection. Several of those involved in the project proposed that we could further our understanding of the interview process by interviewing the interviewers. This additional aspect was original and allowed us to collect information concerning the experiences, insights, and challenges of the interviewers. This new material will be beneficial for future research with vulnerable populations.

Although the interviewers were prepared for the difficulty of hearing the participants' stories, and the mechanisms put in place were, according to the interviewers, very helpful, it became apparent at the end of the study that these mechanisms were too informal. Moreover, although the interviewers were invited to take advantage of debriefings, talk to the interview coordinator, or seek help from one of the trained researchers, there was no requirement that they do so. One lesson learned is that mechanisms should be formalized for everyone and included as an aspect of the interview process.

Because of the nature of the research, the project coordinators were ethically responsible to both the participants and the interviewers. Several challenges were addressed, including the need for in-depth training, debriefing, and support of the interviewers; teaching of self-care to the interviewers; cultural sensitivity training and teaching about diversity; as well as having interviewers who were flexible and non-judgmental. Timely feedback was imperative, and the advice of an Aboriginal Elder throughout was invaluable. It was necessary to have knowledge of services available in various regions so that the participants could be referred to them if necessary. Of central importance was an interview coordinator who provided mentorship, debriefing, organization of material, and a link between the interviewers and the project coordinators. Many of the interviewers were trained therapists, shelter workers, and psychology students, and their training helped them with the interview process. Partnering with the shelters and knowing that they were there to support the interviewers and participants were paramount. One interviewer noted that "The parameters around the study have to be flexible; I remember on a couple occasions having to scramble for time to get the interviews done because in some cases the women are transient, they are mobile. . . ."

Retention is always an issue for any longitudinal project, and our project proved to be no exception. The interviewers, aware of our concerns, made a concerted effort to stay in touch with the participants, expressing passion for the project and compassion for the participants. Their passion was met by that of the participants, who remained in the study because they wanted to help other women, wanted their voices to be heard, and wanted to inform their communities of the services needed.

Two interviewers commented as follows.

[By] staying in, a lot of the women feel that they can help someone else, to inform services for women who might be going through the same thing, or they are giving back or helping in some way; they really stick with it because of that.

I think some of the interviewers have developed really trusting, respectful relationships where the women feel good after they spend time with them. They know that because they say that the women have said, "Gee, I like talking to you, and I enjoy [our time]."

Three of the interviewers commented on the process itself.

Stressful, but we had support. Receiving feedback is important.

I'm committed to the participants to have their voice[s] heard. I have a sense of responsibility to my participants. I've had exceptional participants.

I recognize the need for this kind of data to be collected and wanted to see it through and make sure we have good data.

Participants stuck with the process because they wanted to see change for other women.

They are doing it because they want to see changes for women that are in violent relationships, and so they really feel like they are contributing to something.

Many of the women are staying in this study because they do strongly feel like their voices will potentially be heard.

One strong theme that emerged was ensuring that women had access to services. It was an ethical dilemma when the lack of services became apparent to the interviewers. As one interviewer commented,

> *I know the women's centre, the sexual assault line, they have the twenty-four hour, and they focus on sexual assault, so I have referenced that to people, but I think with the new bereavement centre the grief counselling seems to be something that really needs to be available.*
>
> *I guess the scope of or the lack of services for the women that I have worked with is also thin. One of the areas, you know when we talk about, we talk about shelters and why you can't access them, and then housing, just all of the different issues that come into play, that come along with having an abusive partner and having abusive partners who are in jail and then being released and not knowing when they are being released, and so it's just all of the diverse factors....*

Conclusion

Women who have been abused by their intimate partners are often a hard-to-reach target population and an even harder population to retain in research studies. For many reasons (e.g., safety, stigma, money), they often move and do not want to be "found." This makes it difficult to keep them involved in a research study. This longitudinal study faced ethical dilemmas and ongoing challenges in retaining the participants. Since we are aware of no other longitudinal studies of this kind in Canada, much of what we have learned is first-hand knowledge. Several ethical issues became apparent during the interviews, and as feminist researchers we tried to address them as they arose. We knew that we needed the help of community partners, and they were with us throughout the project. The partners helped with counselling the participants if they needed it, they gave us invaluable advice when we needed it, and they gave us physical places in which to conduct many of the interviews. We learned how important it was to give support to the interviewers throughout the project with debriefing, sharing stories, education, humour, self-care, and appreciation. We had the sage advice of an Elder who talked to the interviewers, helped us with problems that we encountered, and became a major part of our team. "Interviewing the Interviewers" helped to give us more in-depth knowledge of what was happening to the interviewers.

Acknowledgements

We would like to acknowledge the Social Sciences and Humanities Research Council of Canada Community-University Research Alliance Program, which provided funding for "The Healing Journey: A Longitudinal Study of Women Who Have Been Abused by Intimate Partners."

We would also like to acknowledge the Dean's Research Award from the Faculty of Arts, University of Regina, which provided funding for "Interviewing the Interviewers."

We would like to offer our deep appreciation to Luther College, University of Regina; Maria Henrika from Transition House, Regina; Tamara's House, Saskatoon; Deb George from the Domestic Violence Unit at Family Services, Regina; and Meghan Woods and Laura Taylor, project coordinators for the "Healing Journey." In addition, we would like to acknowledge the Prairie Action Foundation for supporting the project.

THE ETHICS OF ENGAGEMENT
LEARNING WITH AN ABORIGINAL COOPERATIVE IN SASKATCHEWAN

Isobel M. Findlay, Clifford Ray, and Maria Basualdo

Introduction

If community-based organizations often feel researched to death, remote Aboriginal communities can feel multiply disadvantaged by research invested in academic priorities and insufficiently sensitive to the unequal distribution of benefit and reward (Smith, 1999, 2005). In the context of a colonial history of research, this chapter shares the Community-University Institute for Social Research's (CUISR) experience of ethical, place-based, collaborative, interdisciplinary research in a four-year partnership with one Aboriginal cooperative: Northern Saskatchewan Trappers Association Cooperative (NSTAC). Working together has revitalized our community-based research (CBR) practice—animating what we call the three Rs of ethical, engaged CBR (research, relationships, and reflexivity). In the process, we are reclaiming from layers of reduction and distortion trapping's rich history as knowledge ecology and sustainable livelihood and reframing individual, community, policy, and programming horizons.

This chapter shares something of our research journey as we worked to build trust and relationships while learning from and celebrating capacity in communities so often targeted as rich sources of data (Smith, 1999) or constructed as "problems" to be solved (Findlay & Wuttunee, 2007; Ponting

& Voyageur, 2005). The chapter tracks how researchers became engaged in research as ceremony (Wilson, 2008), understanding "reciprocity as an ethical starting place" (Kovach, 2009, p. 19), while recentring the knowledge of community members (Smith, 1999). The NSTAC-CUISR research is remapping the territory according to a venerable but adaptive ethics of place. It is retelling stories of traditional trapping's capacity to preserve Aboriginal culture, protect the land, increase economic opportunities, produce sustainable food, engage young people, and reconnect generations. Learning across our differences has yielded place-based learning (Davidson-Hunt & O'Flaherty, 2007), new meaning making, new patterns of identification, and lessons about what such community-based inquiry can mean for social change, sustainability, and justice. If they like other Aboriginal communities have experienced research as theft from away, NSTAC looks to research as a resource consistent with an Elder's advice to start "researching ourselves to life" (as cited in Castellano, 2004, p. 98).

First, this chapter introduces this community-university partnership. Second, it shares the obstacles that we faced and related efforts to rethink CBR and develop our three Rs of ethical community engagement. Third, it considers how the evolving research reframed trapping and engaged us all in a larger trapping vision of preserving Aboriginal culture for hope, healing, and health. And fourth, the chapter reviews lessons learned about the ethics of place and the place of ethics for reconstructed identities, thinking, and action for healthy communities.

Introducing the Community-University Partnership

CUISR

In its first years, CUISR focused on its host city, Saskatoon, before extending its range in 2007 to five interdisciplinary strategies, including the ongoing analysis of community-university partnerships. Authentic partnerships between community and university participants are central to CUISR and reflected in its governance (fifty percent community and fifty percent faculty representation) and founding principles facilitating collaboration and accountable, objective reporting of results (Moote et al., 2001; Sanderson, 2005). Firmly in and for the community, CUISR has built social capital (Coleman, 1988) through research efforts to bridge perceived divides; offer a forum to convene, share ideas, and form coalitions; and provide credible research that can help communities to build capacity, leverage funding, and change policy for the greater good.

Despite CUISR's investment in governance, infrastructure, and researcher training, authentic partnerships at the heart of ethical research remain challenging, especially outside our comfort zone of Saskatoon. Moving into unfamiliar territory (for CUISR) has importantly reshaped our sense of "who we are and how we came to be neighbors and relations in this land" (Castellano, Archibald, & Degagne, 2008, p. 403) and helped to humanize the land reduced to empty space on federal and other maps. In entering the culturescape and ecoscape of northern Saskatchewan, we came face to face with the "Canadian legacy to a people who once governed their own affairs in full self-sufficiency" (Hamilton & Sinclair, 1991, p. 1), with the rewhitening or re-emptying of empty space (Banerjee & Tedmanson, 2010; Panelli, Hubbard, Coombes, & Suchet-Pearson, 2009). We came face to face with the legacy of racism that led "to a vicious circle of dispossession . . . result[ing] in extreme poverty among Indigenous Peoples, which in turn intensifies the racism directed against them. The land problem and the problem of racism must be addressed together; they are the same problem" (Madame Erica Irene Daes, as cited in Venne, 2008, p. 26). Only by addressing them together will we effectively address the racism that is not exceptional but foundational to Western rationality and "progress" (Bhabha, 1994).

NSTAC

Although often represented as in the way of development or out of the way in a desolate wilderness, in eighty fur blocks spanning 500,000 square kilometres across treaty territory (Treaties 6, 8, and 10), 2,400 trappers still live off the land. They continue to teach how to respect nature's gifts, living "the healthiest lifestyle you can have here," according to one trapper (as cited in Pattison & Findlay, 2010, p. 35), and they are willing to share the fruits of the land, as they have for centuries. NSTAC is mandated to monitor and guide trapping, develop policy, deliver training, and lobby government. Just over sixty years after the provincial government in 1946—without consultation and without regard for either natural or traditional boundaries—divided the province into two wildlife management zones, and forty years after forming the northern trappers' association, the northern trappers incorporated as a not-for-profit cooperative in 2007. They did so in efforts to discard colonial legacies and become sustainable, engage youth, and build capacity by regaining traditional knowledge of trapping as integral to the social fabric rather than as peripheral vocation (Nelson, Natcher, & Hickey, 2005). From the standpoint of the government—the primary funder—the restructuring enhanced NSTAC's legitimacy, accountability, and transparency. To ensure legitimacy in the eyes of its members, however, NSTAC needed to communicate the benefits of legal incorporation while respecting the values of its Aboriginal membership (Métis, Cree, and Dene). It needed to engage member wisdom and energy, integrating cooperative and traditional trapper governance from a proud history of a knowledge economy sustaining livelihoods long before the mainstream thought that it had discovered the notion.

This was the focus of the research proposal that NSTAC initiated and the basis of our partnership to re-create a future for Aboriginal youth living the colonial legacy. And the stakes are high. If the trappers cannot prove that they are maintaining a traditional lifestyle on the land, then they open the door to unfettered development without infringing on Aboriginal or treaty rights (Nelson, Natcher, & Hickey, 2005) or incurring the duty to consult (Government of Saskatchewan, 2010a; Newman, 2009). The research process encountered challenges represented by (a) the extent of the territory, travel, and other costs; (b) academic timelines and priorities impacting short-term research engagement; and (c) the legacy of persistently colonial curriculum and pedagogy. The culture of an academic place continued to threaten and thwart the development of ethical relations with the stewards of an Indigenous place.

Facing the Colonial Legacy

Since we were unable to rely on CUISR's profile in the North, a critical agent in building the partnership was an NSTAC friend and adviser to President Clifford Ray, a northerner who knew CUISR personnel and had the benefit of an education on the land and in the university. He participated in early research discussions to prepare the student intern for a six-month internship (part of our student training commitment to SSHRC). The graduate student selected from applications to an advertisement by the research team (though Clifford could not attend the interview) brought experience of cooperative research and development and received training in research ethics and methods, roles and responsibilities of team members, and issues of authorship and dissemination. At our first meeting, Clifford brought rich stories of the meaning of trapping and an armful of maps to highlight Aboriginal markers of meaningfulness erased as much in government maps oriented around government facilities as in educational curricula promoting a world of whiteness, of progress and the pioneer spirit (Tupper & Cappello, 2008). Such erasures replay the doctrine of *terra nullius* that licensed a settler sense of entitlement in the first place (Henderson, Benson, & Findlay, 2000). The maps reminded us of the land's importance for lives and livelihoods. They insisted on our responsibilities to treaties that remain unmarked on provincial maps, though their promises that "the same means of earning a livelihood would continue" were at the heart of the "relationship of reciprocity" understood by Aboriginal signatories and the basis of all our constitutionally protected rights and responsibilities on this land (Talbot, 2009, p. 162).

Focusing in the first phase on governance and engagement, the student completed a literature review, a participant observation, and semi-structured interviews (individual and group) consistent with participatory action research principles (Fals-Borda & Rahman, 1991). We soon discovered the limitations of our approach in communities six and more hours away by road and with uneven access to cellular or e-mail services. The student researcher was able to attend an executive meeting and an annual convention to observe and to meet informally with the cooperative's members. The majority of interviews (lasting from twenty minutes to two and a half hours) with board members, Elders, NSTAC members, women trappers from Manitoba and Alberta, and community members identified by NSTAC president, Clifford Ray, were completed during the River Gathering Festival in Pelican Narrows, 10–13 August 2007. Interviews were thus limited to those who participated in the event and who had the resources and desire to attend it (the majority from the east of the province), and geographic distance between and linguistic diversity of researchers and NSTAC members limited the interaction that one might expect from CBR. With minimal resources for travel, and with a six-month window dictated by student program timelines, we faced considerable odds.

Despite training and supervision by a community researcher and principal investigator experienced in Aboriginal communities, the student continued to understand himself as an observer engaged in data collection and not as a team member learning with and from others. For all his efforts to listen, learn, and capture the social importance of trapping, the trappers' voices were overwhelmed in a first draft report by the dominating narrative of trappers who were dependent on government, poorly prepared, and facing overwhelming odds (global economy, humane trapping standards, and international boycotts). Trapping remained as cut off from its holistic history and current practice as it does in colonial narratives deeply embedded in university "hidden curricula" (Margolis, 2001), media, and government publications.

Government policy responsible for the erosion of Aboriginal, trapping, and treaty rights garnered only a passing reference—as did environmental issues impacting renewed and shared interest in sustainability, food security, and healthy lifestyles. Nor did trapping become visible as a customary practice regulating human behaviour, teaching people their place in the world, their roles, and their responsibilities to "all their relations." It became clear to us that the draft's shortcomings owed much to truncated timelines and our underestimation of the power of colonial narratives of "progress," research associated with distance and disinterest, and training in academic writing that paid scant attention to audience and purpose. With too little ongoing interaction with the trappers, the student marginalized their voices, forgetting whom the report was for, what it aimed to achieve, which actions it might support, and which uses others could make of it. It was a humbling lesson on colonial legacies and the need to be more critically vigilant to decolonize our research practices.

Learning to Decolonize CBR

In their decolonizing efforts, researchers draw on participatory and CBR practices associated with trust building, local knowledge, a type of "revolutionary science" that acknowledges its politics while maintaining its "discipline" (Fals-Borda, 1987, p. 330). They aim to put community participation at the centre of activities designed for policy change; enhanced capacity; enriched discourse, dialogue, and dissemination; and multiple and fluid identities. They have an ethical interest in how people give meaning to their lives, in exploding myths and nourishing stories that help people to imagine alternatives and value their own agency, in seeing how "truths" have been constructed and can be deconstructed and displaced/replaced.

The new research paradigm moves from research on to research by and with communities (DeLemos, 2007), from positivist distance and disinterest to critical inquiry focused on socially and historically constructed power relations and knowledge (Carroll, 2004; Crotty, 2003; Kincheloe & McLaren, 2005). As Boser argues, such community-campus participatory approaches to "co-generating knowledge and, potentially, sharing decision-making based on that knowledge . . . [bring] new sets of social relations for research and . . . *a new set of ethical challenges*" (2006, p. 9; emphasis added). Critical reflection on the "political location and power involved in the social relations" as well as "the potential risks for all constituents" is thus key (pp. 15, 17). Jordan (2003) similarly cautions that, in becoming mainstream, participatory methods are being appropriated by neoliberalism. If Kesby (2005, p. 2047) likewise warns against romanticizing participation that can be a form of tokenism or other tyranny, he also argues (after Foucault) that power has positive valence too and can be transformational and enabling in undoing colonial apparatuses when "disrupted by the reflexivity" of those who "can exercise discretion, innovation, or resistance."

Building on feminist, postcolonial, and participatory research, Maori scholar Smith (1999, 2005) challenges conservative resistances to social justice, recording the extractive and assimilationist history of colonial research embedded in fragmented social sciences and "disinterested" experts monitoring and measuring marginalized populations. In the process, local knowledge was disdained, ignored, or destroyed, while lucrative expertise was legitimated—and justified in turn as the "natural" order of things colonial: encroachment, dispossession, and exploitation of peoples and their resources. This colonial history has produced a "consensual hallucination," a delusion of superiority, or an "epistemology of ignorance" whereby white people "understand the world they themselves have made" (Mills, 1997; Sullivan & Tuana, 2007, pp. 1–2).

The hallucination that allows settler communities to overlook both their own complicity in colonial violence and the achievements of Aboriginal people has proven hard to dislodge yet has stimulated decolonizing strategies and "counter-stories" among Indigenous and like-minded researchers both "humble and humbling" in the "recovery of ourselves, an analysis of colonialism, and a struggle for self-determination" (Smith, 1999, pp. 1, 7). Such research unpacks scholarship's complicity in producing and reproducing inequalities and injustices in white settler society (Findlay, 2003; Razack, 2002), in D. E. Smith's (1990) "relations of ruling," where codes of cultural difference locate responsibilities within othered communities while deflecting attention from mainstream responsibility (Tupper & Cappello, 2008).

In retrieving "the ancient memories of another way of knowing," Smith finds hope for creating space "for alternative imaginings to be voiced, to be sung, and to be heard (again)" (2005, p. 87). In this context, the United Nations and other international, national, regional, and local bodies have been rewriting protocols and ethical guidelines to protect Aboriginal interests (Castellano, 2004; Weir & Wuttunee, 2004; Wiessner & Battiste, 2000) and principles of ownership, control, access, and possession (Schnarch, 2004). When the United Nations asserted Indigenous peoples' right of "control over all research conducted on their people and any aspect of their heritage within their territories," it also stressed that we all have a stake in such controls and protections (Wiessner & Battiste, 2000, p. 387).

Smith (2005) argues for multiple strategies "across multiple sites" for purposeful, transformative research undoing "corporate layers of research" in the interests of a growing Indigenous "research community" (pp. 88–89) concerned to protect against new incursions into "Indigenous knowledge once denied by science as irrational and dogmatic" (p. 93). For Smith, research ethics are fundamentally about "reciprocal and respectful relationships," as in Rigney's "Five Rs: Resources, Reputations, Relationships, Reconciliation and Research" (2002, pp. 27–28; as cited in Smith, p. 97). From Smith's "Community-Up Approach" to researcher conduct, we have learned much about community defining the terms and the spaces of engagement, listening before speaking, the principles of generosity and modesty, and the importance of reflecting on "insider/outsider status" (p. 98). We have learned as much from Wilson's (2008) "research is a life changing ceremony," as from Cora Weber-Pillwax's research as respect, reciprocity, and relationality as well as accountability to "all our relations" (Wilson, 2008, pp. 58–61).

Displacing distance and disinterest as research values, they put at the heart of things Aboriginal peoples, power, knowledge, and experience to unsettle dominant (and dominating) logics. In the process, they reconnect that which modernity uncoupled—thought and knowledge from spiritual, ecological, and social relationships—in but one version of the violence done to Aboriginal peoples. Modernity thus created a cognitive *terra nullius*, as it were, where different ontological, epistemological, territorial, and other orders were reduced to caricatures of themselves or empty shells discarded in favour of colonial calculation, indifference, or caprice. As Supreme Court Justice Abella (2009) argued, "indifference is the incubator of injustice." Indifference to Indigenous stewardship of place is the unethical prelude to colonial injustice.

Developing the Three Rs of Ethical, Engaged CBR

In developing the three Rs, we learned too from the ongoing work of the Indigenous humanities—a strategically labelled and actively produced category mistake or misnaming such as Spivak (1993) welcomes as the limit of authority and the place of progressive change—that unsettles mainstream thinking about what counts as knowledge and expertise (Battiste, Bell, Findlay, Findlay, & Henderson, 2005; Findlay, 2003). Workers in the Indigenous humanities, guided by Indigenous academics and developing new methodologies to decolonize ideas, individuals, and institutions, are committed to perceptual, attitudinal, intellectual, social, and other change. The work is about redeeming expert knowledge, making inquiry more sociable and accountable, and giving more complex accounting of identities and institutions.

While laying claim to the rigour and authority of the traditional humanities, we refuse to relinquish the master's tools (Lorde, 1984) to dominant interests but reshape and redeploy them to promote transformative ends and collective benefits. We hold mainstream institutions to account by rereading economic, cultural, legal, and historical canons; recentring and revaluing Aboriginal knowledge and heritage; attending to multiple and conflicted histories and critical geographies; respecting the authority of Elders and educators, court workers as well as cultural workers; and refusing to be confined by colonial constructions of identity and inequity. Decolonizing is important for all of us because colonialism has taught us negative strategies of difference, habits of hierarchy and deference, and patterns of commodifying and compartmentalizing that rationalize the most irrational acts (Henderson, Benson, & Findlay, 2000).

Thus, demystifying and democratizing research as engagement is importantly related to relationships and reflexivity, to extended timelines and investments in listening, linking, learning, and leveraging together, to challenging systems and stereotypes and changing relationships within and across partnerships. As Wilson (2008, p. 7) argues, "Relationships don't just shape indigenous reality, they are our reality." Indigenous research is about maintaining "accountability to these relationships." Similarly, Kovach (2009, p. 12) underlines our responsibility to go beyond the binaries of Indigenous-settler relations "to construct new, mutual forms of dialogue, research, theory, and action."

Our three Rs, then, are necessary correctives to colonial education's coercive three Rs—"Such shame. Such assault . . . refined under the rule of reading, writing, and arithmetic" (Halfe, 1994, p. 105)—impacting so tragically on Aboriginal youth targeted by residential school policy "to kill the Indian in the child" and the policy's intergenerational effects (RCAP, 1996a, 365). As one trapper put it,

Our young people are losing their culture and tradition and language. ... They want their culture and tradition as an Indian, but then they have to have education to live in this modern day.... They are going towards the whites now, towards the white way of life. And so that is affecting our way of life too. (as cited in Pattison & Findlay, 2010, p. 33)

Another focused on the need to create opportunities:

Not everybody can teach or go into the health training aspect. Yes, it is good to have those people, but those that cannot, must have an alternative.... Maybe they are more comfortable in fishing or trapping. If that is what they are comfortable in, then that is what they like to do, then give them an opportunity to make a living. (as cited in Pattison & Findlay, 2010, pp. 31–32)

Yet another spoke to holistic and healing education in the bush: "Here in the bush you don't have to use a pencil. You have to use your brain because that is your gift to use your brain and your heart" (as cited in Pattison & Findlay, 2010, p. 33).

Instead of the low expectations of the colonial classroom, our three Rs promote heightened expectations and mutual learning, reciprocity, relationship building, and community renewal. Working together, we acknowledge our shared responsibility for actively producing (never simply discovering) data (Schnarch, 2004). In our practice, the three terms—research, relationships, and reflexivity—are now importantly interrelated and critically reflected from first meetings of community partners, student interns, CUISR staff, and faculty to discuss research design and direction until final meetings to approve ongoing dissemination. The research is rigorous and the relationships strong to the extent that we take the time to reflect continually on who we are, how we do what we do, which benefits accrue, and to whom. Ensuring relevancy as a means of "giving back to community" (Kovach, 2009, p. 100) requires attentiveness to difference that listens to learn and that respects the unfamiliar without distorting or domesticating it. That deeper listening, we've learned, is nourished when we avoid taping unless it is the participant's choice and turn to notebooks (or laptops) only after listening and processing together.

The term "reflexivity" (Bourdieu, 1990; Kesby, 2005) signals resistance to and unpacking of academic orthodoxies and monopolies and in turn is resisted more or less vehemently to the extent that it threatens hegemonic structures of privilege and exposes cultural presumptions about protocols and practices. Reflexivity is resisted not only within the academy, which often associates it with endless self-regard (Sultana, 2007), but also among community-based organizations, which associate it with ivory tower introspection and inaction. In contrast, for us reflexivity is an iterative process, a recognition of "the politics of representation" (Kovach, 2009, p. 33). It is an important double gesture of demystification ("research is not rocket science"; Schnarch, 2004, p. 88) and a redirection of the research gaze at our own activities as well as at the policy-programming outcomes of quantitative research that pathologizes Aboriginal peoples (Findlay & Wuttunee, 2007; Salée, 2006). It is also a vital redistribution of authority redefining the research community. It repays time and resource investment despite the attenuated timelines of tenure and promotion and the reporting requirements of funding bodies, not to mention the attention span of political cycles and seasons. It repays in its heightened sense of place and people (and our responsibility to them) and in the visible value that it assigns to the daily rituals and relationships, to the diversity of input in community-building efforts. It is, as Sultana (2007, pp. 374–375) argues too, "integral to conducting ethical research" in its responsiveness to the "contextual, relational, embodied, and politicized" contexts of fieldwork.

So reflexivity and continual revisioning of who we are and what we are doing are at the heart of our research, even if they can be unsettling for those still establishing their identity in the community and their authority as researchers. Sharing the theory of decolonizing research was the first stage in expanding the student researcher's conceptual horizons. The next was to support his growth as a researcher in a six-month mentorship with an Aboriginal undergraduate intern from the community that was the partner in another project. She could share her deep knowledge of her language, culture, and community while benefiting from the graduate student's experience with literature reviews and academic research and writing. The results were gratifying: the Aboriginal student was soon presenting confidently at regional, national, and international conferences, and the graduate intern became more thoroughly immersed in both projects, increasing his own research capacity. As we worked together to disseminate findings in ways useful to communities and respectful of their protocols (Silka, 2003; Wiessner & Battiste, 2000), relationships strengthened, the iterative reflection deepened understandings, and the whole team of researchers found their thinking constantly challenged and combined in a common vision for communities in the North—a positive vision of enormous accomplishment, capacity, and creativity. A truly "life changing ceremony" (Wilson, 2008, p. 61).

We met formally and informally in Pelican Narrows, La Ronge, Saskatoon, and places in between (often Prince Albert), and we attended further conventions, workshops, and conferences and expanded the team's horizons beyond the initial project. Increasingly, Clifford helped to present our research findings to diverse audiences and to design research to investigate ecotourism and justice trapline as parts of the larger goal of cultural revitalization to build futures for youth "to become something, to be proud of something knowing that they have a title," as one trapper put it (as cited in Pattison & Findlay, 2010, p. 34), beyond becoming fodder for the justice system, in which they represent seventy-seven percent of the prison population (Calverley, Cotter, & Halla, 2010). The trapper went on to note that,

> *When they get in trouble, when they break the law, well of course they send them to jail or give them a sentence. But send them to a camp where they can learn about their culture—how to trap, how to hunt, and all that was done in the old days. If they start learning about the Indian people's ways, maybe they can learn about who they are.* (as cited in Pattison & Findlay, 2010, p. 34)

The ceremony of consent was a protracted social process of building relationships, of giving and receiving gifts of food, fun, and friendship long before and after signatures were secured on written consent forms (or oral assent given to readings of the forms).

We engaged more people in the research to reconstitute the relevant community. If the definition of community (who is entitled to speak for and with the trappers) remains an issue for some, our strategy learned from NSTAC to extend community boundaries and welcome all who wished to participate, inviting band and village councils, schools, youth, and Elders. We leveraged resources from a Cooperatives Secretariat Cooperative Development Initiative (CDI) Innovation and Research Grant to investigate ecotourism, including cooperative and business training. Understanding the role of sustaining infrastructure and networks of support, we engaged an Advisory Council from the cooperative sector, government, education, health, tourism, as well as regional and community councils—all of whom became thoroughly invested in the NSTAC vision.

In training sessions, we put into practice our learning about culturally coded practices from conference design to training and research that perpetuate the discursive inequities and insensitivities of the status quo. We learned from witnessing training delivered in the classroom with minimal interaction and community consultations based on PowerPoint slides full of statistics, tables, figures, and technical terms presented by an English speaker to Cree and Dene speakers. We learned from the violence of a talking circle used to engage Aboriginal women while the chair held forth almost without interruption. Learning of the need for proper protocol, community control, indirect styles, and "fertile ground," including the bush, for skills and knowledge to develop (Nelson, Natcher, & Hickey, 2005), we promoted intergenerational dialogue to engage Elders, senior trappers, and youth in the places in the community that mattered to them. A developer told stories of cooperative development; we talked about the NSTAC vision before witnessing cultural memory in action as trappers and Elders told stories that had the youth seeing and articulating anew what they had taken for granted about the North, finding the answers that were always there waiting for them. Just as in participatory action research, there was no manual, only protocols of respect to do this work.

Talking, feasting, and playing together opened eyes and minds to pride in a traditional way of life, to a rich history and powerful teachers. Cooperatives and cooperation were associated like trapping with "good management and accountability," with "working and learning all together for our communities, members, and justice," with "putting community first" and "uniting by alternating leadership." They were associated with "making everybody strong," "families helping each other," and "listening to the Elders and respecting their knowledge." As they talked about cooperatives, they remembered their own proud history of cooperation and found new solidarity in a shared history of principled practice that put people before profits.

We received notice of the CDI award in February 2009 and completed the research project by the end of the federal fiscal year (March 2009). We had solidified relationships and learned to speak each other's language to the extent that we completed everything within a month in mid-trapping season and mid-academic term! While Clifford began to teach us to read natural signs on the land as legible as any city sign once our senses were adjusted, the youth taught us some Cree, and the Elders honoured us as community members. As we challenged notions of legitimate community, we stretched the research community, hiring (with the CDI funding) high school students as researchers in intergenerational dialogue with Elders, using whatever means they chose to record their findings (photography, painting, performance, and storytelling). Therefore, it was less about "discovering" the relevant community than about reconstituting it and recognizing our accountability for interventions and outcomes.

Remapping, Retelling, "Rewriting, and Righting"

Regarding the ethical imperative to rewrite in order to right (Smith, 1999) and to relearn from and on the land for healthy communities and environments, our aspirations are both ambitious and modest: "cognitive justice" and "prudent knowledge for a decent life" (Santos, 2007). Although its history is often reduced to the history of the fur trade (Morton, 1973; Ray, 2005), trapping has an ancient history as a sustainable knowledge economy, though there was no word for "trapping" in Indigenous languages. It is a history celebrated by the Elders as *pimâcihowin* ("making a living") connected to *pimâtisiwn* ("life") and *askiy* ("land") as the source of life and guaranteed by the treaties as a continuing right (Cardinal & Hildebrandt, 2000, p. 43). It represented life as a holistic balance of skills, knowledge, and dependencies linking human survival to sustainable practices and responsible, respectful stewardship of the land.

Elder Peter Waskahat reinforced the powerful legacy deriving from "a way of life based on...our own teachings, our own education," which "taught how to view and respect the land and everything in Creation. Through that the young people were taught how to live, what the Creator's laws were, what were the natural laws, what were those First Nations' Laws" (as cited in Cardinal & Hildebrandt, 2000, p. 6). According to Elder Bart McDonald, "The land is who we are....That was part of our livelihood....The teaching of respect associated with the concept of *pimâcihowin* provided guidance for the ways in which individuals conducted themselves when exercising their duty to provide" (as cited in Cardinal & Hildebrandt, 2000, pp. 46–47).

In the colonial era, one word was replaced by three terms—"hunting," "fishing," and "trapping"—derived from the division of labour and the production of commodities, and all three were under the aegis of progress as predation. In retrieving what was always there to guide, if we could only decolonize our perception and cognition, the Elders, trappers, and youth achieved "the art of the impossible in the realm of the improbable," finding "the collective strength to return to [their] role as the teaching civilization, not the willing learners of modernity" (Henderson, 2008, pp. 10, 48).

In one of many such moves without consultation with Aboriginal peoples, the federal government's 1930 Natural Resources Transfer Agreement transferred natural resources to provincial responsibility, "increasing regulation" over time and eroding trapping rights and ties to a whole way of life (Passelac-Ross, 2005, p. vii). Trapping was effectively reduced from "a unique, social, spiritual, and cultural relationship with the land and its resources" to a "commercial activity" subject to the "same regulatory regime that applies to all trappers, without concern for the Aboriginality of the trapping activity" (Passelac-Ross, 2005, pp. 16, 37).

Wanting to preserve their rights, trappers bought trapline licences and registered traplines, ironically further reducing them to parity with non-Aboriginal trappers in their subjection to paralyzing regulatory restrictions. Similarly, provincial government policy is firmly focused on the interests/incursions of resource companies, while western wildlife management and leisure hunter and fisher values (highly visible in online and print versions of the *2010 Saskatchewan Hunters' and Trappers' Guide* [Government of Saskatchewan, 2010b]) trump and restrict treaty and trapping rights (Calliou, 2000).

Terra nullius legitimates ongoing encroachments, exploitation of the land and its resources, and dispossession of its people; of course, despite jurisdictional and policy blind spots, this land is not empty. As one trapper put it, "There is no unoccupied territory in Saskatchewan: the duty to consult remains." Another recalled that, when "the mace and beaded pillow (made by Florence Highway) help open the Saskatchewan legislature, they are a reminder from the Elders that we have treaties—living, breathing documents—that are part of who we are." It is responsibility to these truths that motivates our ongoing research on *pimâcihowin* to correct the record whereby "Part of the corporate memory of provincial resource management agencies is that Aboriginal and treaty rights do not exist" (RCAP, 1996b, p. 507).

Conclusion

Societies and their institutions require open, dynamic systems to facilitate new ways of thinking and doing. Our collaborative research practice has been for each of us (and our institutions) an important site of learning, relationship and capacity building, identity formation, and renewal. If we do not feel entitled to do this research, we do feel a sense of deep obligation to share our learning about research deficits to resist and redefine colonial research and the policy and programming that depend on that research. We are motivated by the need to change perceptions of trapping perpetuated by colonial research and encourage appreciation for and ethical relations with the stewards of Indigenous places typically caricatured in public discourse about inhumane traps and Western patterns of conspicuous consumption. Our research aims to solidify and sustain relationships within and beyond NSTAC membership, encourage people to act on their interdependence, and recognize a shared interest in change for hope, healing, and health. It promotes trapping as a key means of revitalizing cultural life and the customary practices that regulate human behaviour, teaching people their place in the world, their roles and responsibilities in communities, and the means for maintaining community welfare and spirit and living in healthy, sustainable ways.

A key lesson learned in our research partnership, then, is the ethical imperative of decolonizing research to rewrite to "reright." We learned a painful lesson about the ethics of place and the place of ethics and the damage done by the insufficiently considered culture of academic places, their timelines, training, and investments in disinterest, priorities and protocols, codes and conventions. They have dominated and distorted precisely because they seem so neutral and natural. The three Rs of engaged, ethical research (research, relationships, and reflexivity) proved to be an enabling mechanism, creating space for dialogue and mutual learning, unpacking cultural codes deeply embedded in how we do research, how we train researchers, and how we disseminate findings. The three Rs helped to demystify while democratizing research and respecting ceremonies of earned consent beyond the written form. They helped us to respect storytelling beyond the rituals of peer-reviewed research. They and the student mentoring that they encouraged helped to stretch understandings of data collection as an iterative process based on deep listening (without undue dependency on audiotapes) and attention to daily rituals of place-based meaning making and relationship building. They helped us to take into data analysis and interpretation that which is often discarded as secondary or superfluous: the contexts of social activities, the sites and scenes of community in action and in reflection on births and deaths, risks and renewals, making space for what and who matters to them. They taught student researchers to attend to audiences and purposes in writing and presenting. Intercultural and intergenerational dialogue expanded our sense of community and added to the potential to mobilize knowledge within and beyond the research process. Learning goes well beyond the research project as researchers collaborate to produce a story that is understood and accepted by the community, the university, funders, and policy-makers—and by audiences in local, regional, national, and international settings.

Research governance, infrastructure, training, and good intentions are never enough. Dispersing authority, sharing the power to define, and ensuring mutually beneficial outcomes remain challenging but essential. For CBR to be both ethical and effective, we need more resources of time, money, people, and expertise than many communities (including SSHRC-funded academic ones) can muster. Without CDI funding, for example, we could not (within SSHRC rules) have hired the high school students and recognized and refined their research capacity, and maintaining our research relationship beyond current funding is an ongoing challenge. Most Aboriginal communities in Canada are already stretched by persistently colonial administrative structures that require them constantly to give accounts of themselves and to do so with insufficient resources. We need to change. We need constant acts of translation that satisfy different audiences and levels of bureaucracy. We need to redefine academic priorities of tenure and promotion and the reporting requirements of funding bodies to recognize the longer-term investments required of ethical CBR to produce "deliverables." Still, a lot has been achieved and can be achieved if we do not measure outcomes within the narrow canons of academic success. Our timelines might be longer, but the relationship building, inevitable engagement in community issues, and enhanced community capacity will be more permanent.

Acknowledgements

We wish to acknowledge funding support from the Social Sciences and Humanities Research Council of Canada for Linking, Learning, Leveraging: Social Enterprises, Knowledgeable Economies, and Sustainable Communities, the Northern Ontario, Manitoba, and Saskatchewan Regional Node of the Social Economy Suite, of which this project is a part. We also acknowledge with gratitude funding from the Cooperatives Secretariat

Cooperative Development Initiative Innovation and Research Program to support research for the ecotourism pilot project.

We would like to thank all who agreed to participate in the research, especially the members of the Northern Saskatchewan Trappers Association Cooperative, who so willingly shared their wisdom and welcomed us at their meetings and conventions. We gratefully acknowledge too the contributions of Dave Elliot, formerly of First Nations and Métis Relations, Government of Saskatchewan, and all who contributed to our ongoing discussion, reflection, drafting, and redrafting—all of which became critical to deepened understanding and appreciation of the world of the trappers.

We also acknowledge the support of Lou Hammond Ketilson, principal investigator, Linking, Learning, Leveraging: Social Enterprises, Knowledgeable Economies, and Sustainable Communities; Len Usiskin, community co-director, Social Economy, CUISR; and Louise Clarke and Bill Holden, co-directors, CUISR.

References

Abella, R. (2009). Human rights and history's judgment. Presentation to the Congress of the Canadian Federation for the Humanities and Social Sciences, Carleton University, Ottawa, 28 May.

Banerjee, S., & Tedmanson, D. (2010). Grass burning under our feet: Indigenous enterprise development in a political economy of whiteness. *Management Learning, 41*(2), 147–165.

Battiste, M., Bell, L., Findlay, I. M., Findlay, L. M., & Henderson, J. (S.) Y. (2005). Rethinking place: Animating the Indigenous humanities in education. *Australian Journal of Indigenous Education, 34,* 7–19.

Bhabha, H. K. (1994). *The location of culture.* London: Routledge.

Boser, S. (2006). Ethics and power in community-campus partnerships for research. *Action Research, 4*(1), 9–21.

Bourdieu, P. (1990). *The logic of practice.* Cambridge, UK: Polity Press.

Calliou, B. C. (2000). *Losing the game: Wildlife conservation and the regulation of First Nations hunting in Alberta, 1880–1930.* LLM thesis, University of Alberta, Edmonton.

Calverley, D., Cotter, A., & Halla, W. (2010). Youth custody and community services in Canada, 2008–2009. *Juristat, 30*(1). Statistics Canada Catalogue Number 85-002-X. Ottawa: Minister of Industry.

Cardinal, H., & Hildebrandt, W. (2000). *Treaty elders of Saskatchewan: Our dream is that our peoples will one day be clearly recognized as nations.* Calgary: University of Calgary Press.

Carroll, W. (Ed.). (2004). *Critical strategies for social research.* Toronto: Canadian Scholars' Press.

Castellano, M. B. (2004). Ethics of Aboriginal research. *Journal of Aboriginal Health*, *1*(1), 98–114.

Castellano, M. B., Archibald, L., & Degagne, M. (2008). Conclusion. In *From truth to reconciliation: Transforming the legacy of residential schools* (pp. 403–410). Ottawa: Aboriginal Healing Foundation.

Coleman, J. (1988). Social capital in the creation of human capital. *American Journal of Sociology, 94*, S95–S120.

Crotty, M. (2003). *The foundations of social research: Meaning and perspective in the research process*. London: Sage Publications.

Davidson-Hunt, I. J., & O'Flaherty, R. M. (2007). Researchers, Indigenous peoples, and place-based learning communities. *Society and Natural Resources, 20*(4), 291–305.

DeLemos, J. L. (2007). Community-based participatory research: Changing scientific practice from research on communities to research with and for communities. *Local Environment: The International Journal of Justice and Sustainability, 11*(3), 329–338.

Fals-Borda, O. (1987). The application of participatory action-research in Latin America. *International Sociology, 2*(4), 329–347.

Fals-Borda, O., & Rahman, M. A. (Eds.). (1991). *Action and knowledge: Breaking the monopoly with participatory action research*. London: Intermediate Technology Publication.

Findlay, I. M. (2003). Working for postcolonial legal studies: Working with the Indigenous humanities. *Law, Social Justice, and Global Development (LGD)* (2003-1), special issue on *Postcolonial Legal Studies*, ed. W. W. Pue, http://www2.warwick.ac.uk/.

Findlay, I. M., & Wuttunee, W. (2007). Aboriginal women's community economic development: Measuring and promoting success. IRPP *Choices, 13*(4), 1–26.

Government of Saskatchewan. (2010a). *First Nations and Métis consultation policy framework*. http://www.fnmr.gov.sk.ca/.

Government of Saskatchewan. (2010b). *2010 Saskatchewan hunters' and trappers' guide*. http://www.environment.gov.sk.ca/.

Halfe, L. B. (1994). *Bare bones and feathers*. Moose Jaw: Coteau Books.

Hamilton, A. C., & Sinclair, C. M. (1991). *Report of the Aboriginal Justice Inquiry of Manitoba*. 3 vols. Winnipeg: Province of Manitoba.

Henderson, J. (S.) Y. (2008). *Indigenous diplomacy and the rights of peoples: Achieving UN recognition*. Saskatoon: Purich Publishing.

Henderson, J. (S.) Y., Benson, M. J., & Findlay, I. M. (2000). *Aboriginal tenure in the Constitution of Canada*. Scarborough: Carswell.

Jordan, S. (2003). Who stole my methodology? Co-opting PAR. *Globalisation, Societies, and Education, 1*(2), 185–200.

Kesby, M. (2005). Retheorizing empowerment-through-participation as a performance in space: Beyond tyranny to transformation. *Signs: Journal of Women in Culture and Society, 30*(4), 2037–2065.

Kincheloe, J., & McLaren, P. (2005). Rethinking critical theory and qualitative research. In N. Denzin & Y. Lincoln (Eds.), *The Sage handbook of qualitative research* (3rd ed.) (pp. 303–342). Thousand Oaks, CA: Sage Publications.

Kovach, M. (2009). *Indigenous methodologies: Characteristics, conversations, and contexts.* Toronto: University of Toronto Press.

Lorde, A. (1984). *Sister outsider: Essays and speeches.* Trumansburg, NY: Crossing Press.

Margolis, E. (Ed.). (2001). *The hidden curriculum in higher education.* New York: Routledge.

Mills, C. (1997). *The racial contract.* Ithaca, NY: Cornell University Press.

Moote, M. A., Brown, B. A., Kingsley, E., Lee, S. X., Marshall, S., Voth, D. E., & Walker, G. B. (2001). Process: Redefining relationships. *Journal of Sustainable Forestry, 12*(3–4), 97–116.

Morton, A. S. (1973). *History of the Canadian west to 1870–1871* (2nd ed.). Toronto: University of Toronto Press.

Nelson, M., Natcher, D. C., & Hickey, C. G. (2005). Social and economic barriers to subsistence: Harvesting in a northern Alberta Aboriginal community. *Anthropologica, 47*(2), 289–301.

Newman, D. (2009). *The duty to consult: New relationships with Aboriginal peoples.* Saskatoon: Purich Publishing.

Panelli, R., Hubbard, P., Coombes, B., & Suchet-Pearson, S. (2009). De-centring white ruralities: Ethnic diversity, racialisation, and Indigenous countrysides. *Journal of Rural Studies, 25,* 355–364.

Passelac-Ross, M. M. (2005). *The trapping rights of Aboriginal peoples in northern Alberta.* Occasional Paper 15. Calgary: Canadian Institute of Resources Law. http://dspace.ucalgary.ca/.

Pattison, D., & Findlay, I. M. (2010). *Self-determination in action: The entrepreneurship of the Northern Saskatchewan Trappers Association Cooperative.* Saskatoon: Centre for the Study of Cooperatives, University of Saskatchewan, and CUISR.

Ponting, J. R., & Voyageur, C. J. (2005). Multiple points of light: Grounds for optimism among First Nations in Canada. In D. R. Newhouse, C. J. Voyageur, & D. Beavon (Eds.), *Hidden in plain sight: Contributions of Aboriginal peoples to Canadian identity and culture* (pp. 425–454). Toronto: University of Toronto Press.

Ray, A. J. (2005). *Indians in the fur trade.* With a new introduction. Toronto: University of Toronto Press.

Razack, S. (Ed.). (2002). *Race, space, and the law: Unmapping a white settler society.* Toronto: Between the Lines.

Royal Commission on Aboriginal Peoples (RCAP). (1996a). *Looking forward, looking back.* Vol. 1 of the RCAP report. Ottawa: Minister of Supply and Services Canada.

Royal Commission on Aboriginal Peoples (RCAP). (1996b). *Restructuring the relationship.* Vol. 2 of the RCAP report. Ottawa: Minister of Supply and Services Canada.

Salée, D. (2006). Quality of life of Aboriginal people in Canada: An analysis of current research. IRPP *Choices, 12* (6), 1–38.

Sanderson, K. (2005). *Partnering to build capacity and connections in the community.* Saskatoon: CUISR.

Santos, B. de S. (Ed.). (2007). *Cognitive justice in a global world: Prudent knowledges for a decent life.* Lanham: Lexington Books.

Schnarch, B. (2004). Ownership, control, access, and possession (OCAP) or self-determination applied to research: A critical analysis of contemporary First Nations research and some options for First Nations communities. *Journal of Aboriginal Health, 1*(1), 80–95.

Silka, L. (2003). Community repositories of knowledge: A tool to make sure research pays off for university partners. *Connection: The New England Board of Higher Education*, (Spring), 61–63.

Smith, D. E. (1990). *Texts, facts, and femininity: Exploring the relations of ruling.* London: Routledge.

Smith, L. (1999). *Decolonizing methodologies: Research and Indigenous peoples.* London: Zed Books.

Smith, L. (2005). On tricky ground: Researching the Native in the age of uncertainty. In N. K. Denzin & Y. S. Lincoln (Eds.), *The Sage handbook of qualitative research* (3rd ed.) (pp. 85–108). Thousand Oaks, CA: Sage.

Spivak, G. C. (1993). *Outside in the teaching machine.* New York: Routledge.

Sullivan, S., & Tuana, N. (2007). *Race and epistemologies of ignorance.* Albany: SUNY Press.

Sultana, F. (2007). Reflexivity, positionality, and participatory ethics: Negotiating fieldwork dilemmas in international research. *ACME: An International E-Journal for Critical Geographies, 6*(3), 374–385.

Talbot, R. J. (2009). *Negotiating the numbered treaties: An intellectual and political biography of William Morris.* Saskatoon: Purich Publishing.

Tupper, J. A., & Cappello, J. (2008). Teaching treaties as (un)usual narratives: Disrupting the curricular commonsense. *Curriculum Inquiry, 38*(5), 559–578.

Venne, S. (2008). Land rights of Indigenous peoples—not racist. *Directions, 5*(1), 25–28.

Weir, W., & Wuttunee, W. (2004). Respectful research in Aboriginal communities and institutions in Canada. In B. Fairbairn & N. Russell (Eds.), *Cooperative membership and globalization: New directions in research and practice* (pp. 207–236). Saskatoon: Centre for the Study of Cooperatives, University of Saskatchewan.

Wiessner, S., & Battiste, M. (2000). The 2000 revision of the United Nations draft principles and guidelines on the protection of the heritage of Indigenous peoples. *St. Thomas Law Review, 13*(1), 383–414.

Wilson, S. (2008). *Research is ceremony: Indigenous research methods.* Halifax: Fernwood.

2
ADVOCACY AND COMMUNITY-BASED RESEARCH

COMMUNITY-BASED RESEARCH AND ADVOCACY FOR CHANGE
CRITICAL REFLECTIONS ABOUT DELICATE PROCESSES OF INCLUSION/EXCLUSION

Gloria DeSantis

> *Splendid innovations can come out of things that do not fit in.*
> —*Chai Chu Thompson, 1991*

Prologue

This chapter is a formulation of my current but temporary critical reflections. It is a case study about an initiative regarding which I have done much thinking and critiquing due to some post-project discomfort that lingers almost two decades later. It is also based on my twenty years of involvement in a variety of community-based participatory action research (CBPAR) projects as a practitioner in the non-profit social service system and on the many conversations that I have had with CBPAR enthusiasts. These reflections are viewed through my university teaching and research experiences over the past several years as a Ph.D. student and now as a research associate at the Saskatchewan Population Health and Evaluation Research Unit (SPHERU). Thus, the lens through which I reflect and write this chapter is that of a "pracademic," wherein I struggle as both a practitioner and an academic to "bring real life experience to the testing and correcting of theory (i.e., a reality check)" (Macduff & Netting, 2010, p. 45).

I acknowledge that the involvement of the project's Steering Group members in writing about our experiences together would have yielded a richer product. However, twenty years have passed since the Steering Group last met. Attempts to find and connect with many of these members were not successful; the two members that I reached did not want to participate because they said they did not have time. Thus, I relied on our Steering Group's meeting minutes, draft documents, and letters to government staff, elected officials, academics, and other organizations to refresh my memory about salient issues.

Finally, as this manuscript went through the editing process, peer reviewers suggested that I make explicit the roles that I played in this initiative. This reflective exercise led me to a deeper understanding of CBPAR and what we might be called to do in this kind of work. I played a multiplicity of roles—as did others in the Steering Group—to have this CBPAR achieve the successful outcomes that it did. I was a facilitator of small and large group community meetings; a data finder and number cruncher; an analyzer; a translator between different constituencies given the different languages they spoke (e.g., governments versus community members); a mediator between the Steering Group and various other constituencies (i.e., the funder, mainstream non-profit organizations, governments) as well as within the group (i.e., due to some of the historical frictions among diverse/racial groups); a research methods teacher; an advocate and defender of the process; a logistics coordinator (e.g., events, meetings); a writer/editor; a secretary at meetings to create records; and, finally, a nurturer as people needed supports and others began to burn out. I now see that juggling and balancing these multiple roles over time has become one of my greatest sources of learning. I hope that they make me a better pracademic.

Introduction

Community-based research and advocacy for change are inherently connected. CBPAR, one of many community-based research approaches (Minkler & Wallerstein, 2003; Reason & Bradbury, 2008), provides an opportunity to explore action and advocacy as part of the research process. The following pages explore a CBPAR case study involving diverse racial and cultural groups' access to the social service system in an urban area of Canada. It is a story about a group of community residents who took control and ownership of a CBPAR initiative and invited academics, non-profit social service organizations, governments, and funders to collaborate at various points in their process. This group of residents, who became the CBPAR Steering Group, were committed to both research and advocacy for change. During their community-based process, they balanced and facilitated an inclusion/exclusion dynamic among different constituencies.

In general, CBPAR is both a philosophy and an approach (Green et al., 1995). As a philosophy, it is located in critical theory, which maintains a spotlight on power relations (Crotty, 2003). It is not a neutral procedure but a value-laden exercise based on a model of research *with* communities, not research *about* communities. CBPAR accepts myriad ways of mutual knowing and learning, and it is values-oriented toward resolving inequities (Israel, Schulz, Parker, & Becker, 1998; Reason & Bradbury, 2008).

As an approach, CBPAR is multi-faceted, messy, and non-linear. It integrates social investigation that creates and exposes new knowledge, learning, and skill building for participants and action to foster change based on the research results (Wallerstein & Duran, 2003). CBPAR is described as a set of practices in response to a person's need to act and seeks collaborative engagement among people (Israel et al., 1998; Reason & Bradbury, 2008). It is best characterized as a spiral of steps (Reason & Bradbury, 2008) in fluid and emergent exercises (Lincoln & Guba, 1985) that responds to a person's need to act on pressing issues (Reason & Bradbury, 2008) wherein oppressed people play a significant, active role (Maguire, 1987; Rahman, 2008). For my purpose here, CBPAR is my preferred label because my exploration of the concept draws *participation* and *action* components to the fore (Reitsma-Street & Brown, 2004). I offer reflections on three additional concepts—*power, voice,* and *knowledge*—that weave their way throughout these processes (Wallerstein & Duran, 2003) and have implications for delicate processes of inclusion/exclusion.

Inclusion/exclusion refers to the movement of various groups of people or constituencies (e.g., marginalized groups, governments, social service organizations, funders) in and out of various research activities. In CBPAR, there are numerous types and levels of participation. Essentially, who participates, to what degree, with what kind of power, when, how, and why, define the extent to which a process is participatory.

I will briefly explore these concepts, beginning with *participation.* CBPAR is biased in favour of the active participation of dominated, exploited, and ignored individuals and groups throughout a process (Reid,

2004a). Some of its processes espouse the active engagement of marginalized people, governments, academics, and funders sitting around the same table, but true collaboration and deep involvement of marginalized people might not occur because the culture of meetings, the implicit power hierarchy, and the attitudes of the people involved inhibit participation (Reason & Bradbury, 2008; Wallerstein & Duran, 2003).

Let us now consider *action* within the context of CBPAR, which is action and change oriented in favour of oppressed groups (Boyd, 2008). The element of action gives groups motivation to get the research done; they know that it is not "just another research project" but something concrete that directly affects them and is expected to change by the end. Action is directed outside the group toward "abolishing unjust policies or constructing new ideas and structures" (Reitsma-Street & Brown, 2004, p. 305) and is conceived as a form of civic engagement (Boris & Mosher-Williams, 1998; Salamon & Lessans Geller, 2008). Advocacy is a process of speaking out about a situation that is unfair, disseminating information, and encouraging change in "individual behaviour or opinion, corporate conduct, or public policy and law" (Rektor, 2002, p. 1; Wight-Felske, 2003), and it is usually directed at governments and others in positions of influence (McCubbin, Labonte, & Dallaire, 2001; Stienstra, 2003). This advocacy can occur at any time throughout the CBPAR process, not just at the end when study recommendations are released.

Let us now consider *power*, how it can be manifested, and as a result who has voice during CBPAR processes. Power is a salient variable in the relations among governments, funders, non-profit organizations (NPOs), and marginalized people engaged in community-based processes. Grabb (2007, p. 211) defines power as "a differential capacity to command resources, which gives rise to structured, asymmetric relations of domination and subordination among social actors." Dominant groups have the power to construct "self-evident descriptions of social reality," which then become common sense (Fraser, 1997, p. 153). Lack of power and control among "research subjects," who often have lower social status than researchers, inhibits participation (Wallerstein, 1999, p. 40) and becomes exclusionary. However, power can also be a positive force in society (Grabb, 2007). As a positive force, it can be productive and includes the capacity to involve marginalized people in rethinking their roles (Kincheloe & McLaren, 2005) and engaging them in collective action. In sum, CBPAR can be a site of both oppressive and productive aspects of power.

Finally, power is reflected in who creates *knowledge* and what counts as knowledge—essentially, whose voices are heard and converted into action during CBPAR processes? The production of knowledge is a social practice (Sayer, 1992). Foucault theorizes that knowledge is power (Farganis, 2004) and that the exercise of power legitimates what is true and what is real (Gordon, 1980). Those who control research (e.g., universities, governments) also control knowledge making (Guba & Lincoln, 2005). Experiential knowledge can become a form of citizen power when people with lived experiences, especially those who are marginalized, speak out.

The diverse racial/cultural groups' case study and my reflections are most closely aligned with a Freirian emancipatory tradition. In this tradition, the oppressed, not the experts, are agents of change (Freire, 2003; Wallerstein & Duran, 2003). Citizens undertake both the conceptual and the practical work, while academics, governments, and social service providers are outside the centre of the group (O'Connor & Williams, 1994).[1] This Freirian approach assumes "that the same social factors that shape relations outside the project will also impact the project"; thus, this approach is carefully designed to minimize these effects (Boyd, 2008, p. 31). This approach differs from "collaborative utilization-focused research," which incorporates rational decision making and cooperative inquiry involving governments, academics, and community members (Wallerstein & Duran, 2003, p. 28; see also Kirby, Greaves, & Reid, 2006). In summary, participation, advocacy for change, power and voice, and knowledge generation were salient concepts in this case study.

Background

The project began when a non-profit organization, whose mandate was community-based research and social planning, was contacted by regional government and asked to undertake a funded study on diverse racial/

---

1 The McMaster Research Centre for the Promotion of Women's Health (O'Connor & Williams, 1994) documents four models of doing research along a continuum. At one end of the continuum is citizen-directed participatory research (most closely aligned with the Freirian tradition), followed by facilitative/educational participatory research, then academic-directed participatory research, and finally traditional research. Each model reflects differing degrees of control held by different participants during different parts of the research process.

cultural groups' access to the social service system.[2] It took place in a mid-size Canadian city with a population of approximately 300,000. The term "racial/cultural group" was chosen because it was inclusive and evoked the best image of a variety of the visible and invisible characteristics of the communities involved in the research.[3]

The non-profit organization (NPO) became the "facilitator organization" (O'Connor & Williams, 1994) (hereafter referred to as the facilitator-NPO) and my employer at the time. The facilitator-NPO chose a CBPAR method based on recommendations from its standing committees—the Research Advisory Committee and the Community Development Advisory Committee. Both committees comprised experts in their respective fields (e.g., university faculty and senior NPO staff). The facilitator-NPO routinely encouraged the blending of research and community development in its work. It promoted community development as empowering processes for "building social relationships, communication networks, and volunteer capacities" (Brennan, 2007, p. 7).

The government funder's research objectives were to

a. review and highlight existing reports on the topic of access;
b. develop a profile of the diverse racial/cultural population;
c. develop a profile of staff, programs available, and people served in existing mainstream and racial/cultural social services;
d. identify barriers to social services;
e. identify goals necessary to reduce these barriers; and
f. prioritize the goals and develop consensus around a model(s).

· · · · · · · · ·

2 The social service system included a wide range of services offered by both government departments and the non-profit sector, including employment and retraining services, child welfare services, family counselling, shelters for victims of intimate partner violence, housing services, income assistance, settlement and integration services, and English-language classes.
3 The Steering Group had a number of conversations during the first few meetings about the term "racial/cultural groups." Statistics Canada definitions of ethnicity, race, mother tongue, home language, citizenship, nationality, place of birth, immigrant, and refugee were brought to the meetings and debated. The group also deliberated over whether First Nations and Métis peoples should be included in the CBPAR project. The group decided that it should be inclusive; thus, the research included First Nations peoples, Métis peoples, immigrants, refugees, and those who were born in Canada but would be considered members of visible minority groups.

The facilitator-NPO's philosophy of doing CBPAR shaped the approach to meeting the funder's research objectives. For example, instead of simply doing mail surveys, the facilitator-NPO wanted community residents to meet together to discuss issues; thus, both mail surveys and discussion groups were organized.

The formation of a diverse racial/cultural steering group was deemed to be the most important first step. I worked in collaboration with two of the facilitator-NPO's board members—one was an immigrant, and one was born in Canada but was an active member of her visible minority community—and embarked on an outreach process to find committee members through faith groups, ethnic associations, and non-profit organizations. The two major criteria for Steering Group membership were individuals who had lived experiences with the social service system and were active volunteers in their racial/cultural communities (e.g., volunteers on boards of directors). After three months of advertising and phone conversations, an initial group of fifteen people met; however, only eleven remained active for the duration of the project, and they became the Steering Group.

Research Journey

The key mechanism to move the initiative forward was the Steering Group. During their first few meetings, they discussed their assumptions, values, dreams, and expectations about the group, community processes, and desired changes. They talked about having the capacity to be knowledge generators and were aware that along with knowledge generation comes action. They also revisited their Steering Group membership in terms of whom they should include and whom they should exclude. There were clear tensions over inclusion/exclusion, as evidenced by the many arguments that ensued. Minkler et al. (2002) experienced similar tensions in their CBPAR study and decided on a team comprising solely people with disabilities. The diverse racial/cultural Steering Group finally decided that only people with lived experiences of the social service system who were also members of diverse racial/cultural communities would be members of the group. Given the diversity inherent in communities, the many potential constituencies (e.g., non-profit organizations, governments), and the limits to the sizes of steering groups, many people were excluded from the nucleus of this project.

I was the only white, non-immigrant person who participated in monthly meetings and worked with the Steering Group. As these meetings and conversations happened, I felt uncertainty and discomfort. On the one hand, I understood their central arguments about wanting to control the initiative because of the power, control, domination, and voice issues cited above. On the other hand, I believed strongly in collaborative, multi-constituency approaches to resolving community issues. When the group decided unanimously to take full control of the research project, I found myself having to let go and trust their wisdom.

The Steering Group was the nucleus of the CBPAR. I sensed that the group wanted to maintain total control and decision-making authority over both the macro- and the micro-elements of the research. However, over time it became clear that the group placed certain constituencies in more trusted places than others and thus was not so exclusionary. For example,

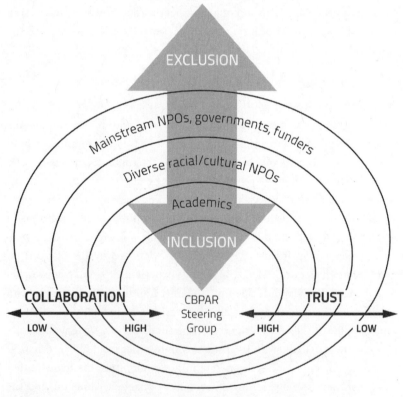

Figure 1. Steering group trust, ease of collaboration, and inclusion/exclusion of constituencies

they regularly sought research advice from the facilitator-NPO's advisory committees comprising academics; these academics were resource people invited to participate in Steering Group meetings when certain tasks were on the agenda (e.g., how to design a phone survey, how to organize focus groups to collect data). Collaboration on various tasks with certain constituencies was easier than with others. The Steering Group also seemed to trust diverse racial/cultural NPOs more than mainstream NPOs, governments, and funders. Figure 1 depicts my sense of the different levels of trust and ease of collaboration in concentric circles radiating out from the nucleus. The figure also reflects degrees of inclusion/exclusion of various constituencies.

The Steering Group deliberated over, made choices on, and implemented a variety of research activities. The following list shows the tasks that they undertook following the citizen-directed participatory approach (O'Connor & Williams, 1994). Academics provided technical assistance with some of these tasks. The Steering Group

- discussed and analyzed census and immigration/refugee data;
- discussed literature on social service access issues;
- designed phone and mail-out surveys to social service NPOS based on examples from other cities and analyzed collected data;
- did outreach into racial/cultural communities across the city to get them involved and with governments and NPOs to get them thinking about issues;
- created written surveys for social service users and analyzed data;
- designed separate workshops for diverse racial/cultural groups and social service providers for data collection[4] and networking;
- found and organized interpreters for focus groups;
- facilitated and recorded focus group discussions and analyzed data;
- assisted in the writing of a discussion document that was circulated around the community for three months to generate conversation and debate;

- - - - - - - - - -

4 Steering Group members were concerned that citizens would not speak freely if service providers such as NPOs and governments were present. In group work, there is an implied hierarchy of "authority" or power (Morgan, 1997, p. 37) wherein governments are perceived to have the most power and marginalized people have the least. As well, the literature shows that different groups can have very different perspectives on the same issue (Krueger, 1994; Morgan, 1997).

- helped to organize follow-up conversations in focus groups[5] to discuss the document;[6]
- formulated priorities and recommendations a few months later; and
- formed an implementation team that comprised some Steering Group and some new members.

Actual Outcomes

This section provides an overview of some of the outcomes of this CBPAR case study. Outcomes are the benefits or changes for individuals or populations from an intervention and indicate *what* has changed; they thus differ from outputs (Hatry, van Houten, Plantz, & Greenway, 1996). Described here are results from the Steering Group's advocacy work specifically. Outcomes were evident in a number of spheres (e.g., individual, community, government institutions, NPOs), but I focus on outcomes for governments and NPOs for the sake of brevity. The outcomes relate to client forms, grant forms, programs, policies, governance, and systems, and they are presented in Table 1. These outcomes occurred within the first two years of completion of the research.

There are examples of longer-term outcomes too. For example, over a five-year period after the CBPAR was completed, other research was pursued by the regional government's Community Services Department regarding elderly immigrants, the local college and university began conversations about creating new courses for different racial/cultural communities modelled after the Native Community Care Program, and the Committee for Survivors of Torture came forward and asked for assistance in its CBPAR work for a neighbourhood-based English as a Second Language (ESL) program with attendant child care.

• • • • • • • • •

5 The central feature of focus groups "is their explicit use of group interaction to produce data and insights" (Morgan, 1997, p. 2). Focus groups also provided a feedback loop for participants to verify the contents of the draft report (Lincoln & Guba, 1985).
6 Focus group participants sought clarification on a number of issues, suggested rewording some of the recommendations, and then spent most of the time talking about how to implement the recommendations; this is an action/advocacy component that began before the report was finalized.

Sphere	Social Service NPOs *(client data collection forms and procedures, specific programs, NPO board governance, the NPO system)*	Governments at Local, Provincial, and Federal Levels *(NPO client data summary forms, grant forms, policies, new planning models, and new service delivery models)*
Initial Outcomes	• A seniors' NPO redesigned its intake forms so that it could better capture and track the diversity of clients that it served. • A family service NPO translated its service brochures into a variety of different languages and enhanced its outreach efforts. • The board of directors of a women's NPO underwent sensitivity training, and the NPO formulated staff and volunteer recruitment equity policies based on the racial/cultural diversity of the city. • The facilitator-NPO formed a Race Relations Committee to undertake organizational assessment and facilitate internal policy change (e.g., equity-based volunteer and staff recruitment policies). • At a systems level, the Immigrant Serving Interagency Network saw its membership increase, its number of meetings increase, and more collaboration among immigrant-serving NPOs and mainstream NPOs, with the goal to maximize people's access to the social service system. • Also at a systems level, a newly formed, community-wide Cultural Interpreters Committee went on to develop a process and a model to serve both clients and NPOs because the research showed a major need for cultural interpreters to serve NPOs, hospitals, and the courts.	• The regional government changed its social service NPO grant application forms to require enhanced data collection and tracking of a number of client racial/cultural/language variables. • The regional government also developed new equity policies requiring NPOs to describe their client bases and explain their fit with the larger community. • A new list of funding criteria included in a recommendation endorsed by the CBPAR process was adopted and implemented by the regional government and used to make more equitable social service funding decisions across population groups and NPOs. • The health council included racial/cultural groups in its mental health planning terms of reference and ensuing planning process. • The federal and provincial governments responsible for settlement and integration programming had watched our CBPAR process from the sideline and approached us to do a CBPAR follow-up study. The ensuing ten-month CBPAR process ended with the recommendation for a new social service delivery model, which they then co-funded. That service delivery model received annual revenues of approximately $3 million to serve clients.

Table 1. Initial outcomes within social service non-profit organizations and governments.

Lessons Learned

This section describes some of the lessons that I learned based on critical reflections of my work on this and other case studies as well as my recent academic research. The focus is on the major theme of inclusion/exclusion and implications for collaboration between CBPAR steering groups and their constituencies. I refer specifically to the key concepts of participation, advocacy and action, power and voice, and knowledge making introduced earlier. In sum, I learned that, depending on the community group, its environment, and the issues, one needs intentionally and carefully to define and facilitate delicate processes of inclusion/exclusion during a CBPAR project.

Participation

Participation is a complex concept. Asking who participates, when, how, and to what degree helps us to see this complexity better. Forming a CBPAR study team, seeing how the members interact with their communities while collecting data, and recognizing the dynamic nature of relationships and collaborations during the process help to illustrate this complexity. Each will be addressed below.

First, the Steering Group comprised people from a variety of racial/cultural communities who had lived experiences with the social service system. The group formed the nucleus of the CBPAR; everyone else was excluded from participating in project governance. The process for selecting people with lived experiences to join the Steering Group was challenging. We began with the premise of finding community "representatives" but decided to invite people based on "their lived experiences." Boyd (2008) notes that you cannot pick one person to "represent" an entire group. For example, in her research with Downtown Eastside Vancouver residents, she clearly heard that "you can't pick one woman to represent all the Indians...or to represent all the drug addicts...or to cover all the gay and lesbian [and] transsexual [residents]...and prostitutes" (p. 30). Other CBPAR initiatives have engaged in similar struggles, including one that saw angry confrontations over legitimate and illegitimate exclusion (Minkler et al., 2002).

> **Lesson 1:** Invite people to join steering groups based on their "lived experiences," not as "representatives" of their communities, and ensure that the group works carefully to build a credible presence during the project.

Lesson 2: Be clear about and be prepared to explain the CBPAR model chosen and its implications for the design and membership of the steering group.

Second, when CBPAR projects are initiated, there are many roles and activities that members of steering groups undertake, but there are many hidden impacts of the choices to participate. For example, Steering Group members facilitated some public meetings and focus groups, but some residents did not want to talk about certain issues because they did not want the facilitator to know about particular problems (e.g., they were involved in a family counselling service). Yet, in another focus group, other residents felt comfortable knowing that someone from their community was taking notes and guarding their data.

Lesson 3: Assess community dynamics and potential steering group members' roles before holding public or focus group meetings in order to ensure that steering group members are appropriately included in various activities while maximizing data integrity.

Third, in this case study, relationships among constituencies and collaborations were dynamic. There were numerous ebbs and flows over time. Others have noted that relationships are constantly negotiated and renegotiated over the course of research initiatives (Kirby et al., 2006). Collaboration also differs across research topics, groups, and stages of the research process (Boyd, 2008). Some of these relationships were intentional, while others just happened. For example, at the beginning of the project, the funder wanted to attend meetings, but the Steering Group did not want it there, so we had to convince the funder that, given the community-directed nature of the project, it had to be excluded from monthly meetings. The funder never really seemed to accept this; constant tension hung in the air as we moved on with our research work without the funder attending any Steering Group meetings. During the research design phase, academics participated in many of the Steering Group meetings. During the data collection phase, Steering Group members themselves and other community volunteers participated in translation and facilitation work at public meetings. During the public meetings to discuss the draft document and recommendations, social service providers were encouraged to attend and participate in a collective critique because it was believed that the more they understood the issues, the greater the likelihood that they would adopt recommendations directed at them.

> **Lesson 4:** Carefully define and facilitate inclusion/exclusion over the course of the project. It is important to be deliberate and transparent about who is excluded from which stages of the CBPAR project.

> **Lesson 5:** Think carefully about choosing someone who has the skill base to facilitate both the macro- and the micro-participation elements in terms of who is included and who is excluded and when.

Advocacy and Action

The research element of CBPAR usually receives funding, but the action element (i.e., advocacy) seldom receives funding. "Research grants are geared towards identifying and assessing needs rather than providing material means to create social change" (Boyd, 2008, p. 40). Thus, advocacy often gets shortchanged. Our Steering Group seemed to recognize that this imbalance was an issue, and we started conversations about advocacy early in the research process. We agreed to meet with governments and non-profit organizations to explain the CBPAR process and the focus on barriers to the system long before public meetings and the release of the draft report. We prepared the ground for what was to come.

Later in the CBPAR process, when recommendations and a final report were released, constituencies and their networks were salient. If the earlier phases of the process had not included academics, governments, non-profit organizations, and funders, then widespread support for recommended change would have been unlikely. In one instance, a federal government department was unintentionally omitted from a public meeting, and recommendations directed at the department were very difficult to implement. In addition, questions about who should play leadership roles to ensure that recommendations were adopted required answers. In the end, what shocked us was that this CBPAR became a stepping stone for the Steering Group to go on to a much larger CBPAR initiative that included developmental research to create a new social service delivery model; this model was later accepted and funded by the provincial and federal governments and became a $3 million service delivery organization.

Lesson 6: Advocacy for change can happen at any stage of the CBPAR process. The lines blur between research activities and advocacy activities when community participation is a central component. The shift from social investigation to social change activity is not tidy. The greater the number of conversations in the community during a CBPAR initiative, the greater the opportunity to build awareness of key issues and the greater the pressure base for change.

Lesson 7: If some groups and organizations are excluded from the knowledge generation phase (i.e., data collection, analysis, and report writing), then advocacy for the implementation of recommendations can be tougher.

Power and Voice

Power, voice, and participation are intricately woven together in CBPAR processes. Some scholars explain that despite their efforts to equalize power imbalances and encourage voice in their community-based processes there were failures (Fay, 2003; Reid, 2004b; Wallerstein, 1999). For example, people deferred to academics, governments, and social service providers during public meetings. It does not always work to check your power and privilege at the door (Fay, 2003). In the present case study, the nucleus of power, participation, and voice was dominated by people with lived experiences. The Steering Group created moments of strategic collaboration when it invited resource people to certain meetings. Thus, there did not seem to be contentious power dynamics requiring amelioration because people were clear about the space in which they were working together. For example, when academics were invited to monthly Steering Group meetings, members of the group seemed to know and accept that they were in a learning space. They did not seem to be fearful about a lack of power and control, and academics seemed to be comfortable with this type of collaboration. As Kirby et al. (2006) state, we must carefully consider our locations and labels, our life experiences, and the types of power that we bring to CBPAR processes.

There are many different forms of power exercised by many different constituencies in CBPAR processes. Some constituencies not in the nucleus of control held by the Steering Group attempted to control the research agenda at various times. First, the funder tried to exercise power over the process using the threat of cancelling the funding contract. Second, social service providers attempted to control the framing of the recommendations included in the final document by dominating public meetings because they wanted to position themselves better to access scarce financial resources. Third, the facilitator-NPO had mainstream organizational power, which it had to use carefully in differential ways.[7] And fourth, when it came to advocacy for the recommendations, certain non-profit organizations not involved in earlier phases of the CBPAR that had the power to alter certain programs simply closed the door on conversations with the implementation team.

> **Lesson 8:** Be aware that there are many different forms of power exercised by different constituencies at different stages of research processes.

> **Lesson 9:** Recognize the labels we wear and that the power that comes with certain labels excludes and silences some but not others.

> **Lesson 10:** If a steering group includes solely people with lived experiences, then invite resource people (e.g., academics) at strategic points along the way to assist with various tasks, but ensure that the steering group maintains control. Create clearly defined spaces in which academics can work with steering group members on research techniques, tools, and processes.

Knowledge Making

Knowledge making began almost at the moment the Steering Group first met and happened at various points during the process. First, when a group of people meet to discuss issues, they unearth new insights and relationships among issues (Kamberelis & Dimitriadis, 2005). Second, during the research, the Steering Group wanted to include the wider community, service providers, and governments in public conversations about draft results, draft documents, and draft recommendations. However, with the

----

[7] At some points during the process, the facilitator-NPO had to work hard to discard its mainstream and status quo image and instead align itself with marginalized racial/cultural communities; at other times, it had to use its mainstream power to prevent other powerful organizations from trying to take over parts of the process.

Steering Group's attempt to be inclusive came challenges that could not be reconciled. For example, NPO workshops and community resident workshops ended with very different results; NPOs did not see their discriminatory behaviour, but their service users did. Academics assisted the Steering Group in its struggle with these common and divergent themes during analysis. Academics encouraged respectful ways of knowing. In the end, they supported the Steering Group in its creation of two distinct lists of recommendations and rationales: one list from the perspective of service users, one list from the perspective of NPOs. With the support of these academics, the finished document and recommendations appeared credible when released.

> **Lesson 11:** Co-construction of knowledge might not be possible or desirable in some initiatives. Be prepared to discern how to rationalize and present results from different sources that simply cannot be reconciled.

Conclusion

This CBPAR initiative brought together people through explicit conversations clarifying that it was more than "just another research project." It was also about advocating for change. Members of the Steering Group saw the research as groundwork for their advocacy work. Their priority was to change the social service system to make it more accessible to people of diverse racial/cultural backgrounds. They brought a sense of urgency to the initiative because they could not as "easily escape spatial, material and historical realities at the end of the day" (Boyd, 2008, p. 29) as do many academics, governments, and non-profit organizations. Interestingly, academics, governments, funders, and non-profit organizations seemed to accept and respect the decisions of the Steering Group.

The group did not complete the research and create change alone, though. The members came together as a group with shared life experiences, directed the CBPAR initiative, and engaged other constituencies to participate in various parts of the process along the way. They actively and explicitly chose who would participate in which aspects of the project. They seemed to be aware of the implicit hierarchy of power that silences some people but not others, even when heterogeneous groups think that they have created a safe and equitable space in which to do research together (see, e.g., Wallerstein, 1999). Some people cannot check their power and privilege at the door (Fay, 2003). Indeed, exclusion of academics, non-profit organizations, governments, and funders from the nucleus of a CBPAR project and their inclusion in other aspects were ways to deal with the implicit power hierarchy that silences marginalized people in many processes (Boyd, 2008; Fay, 2003; Reid, 2004b; Wallerstein, 1999). The Steering Group spent many hours carefully defining and balancing inclusion and exclusion based on the fundamental goal of wanting to maximize the participation, voice, and knowledge-making capacity of marginalized people. The group created and controlled delicate processes of inclusion/exclusion throughout the CBPAR project.

References

Boris, E., & Mosher-Williams, R. (1998). Non-profit advocacy organizations: Assessing the definitions, classifications, and data. *Non-Profit and Voluntary Sector Quarterly, 27*(4), 488–506.

Boyd, S. (2008). Community-based research in the Downtown Eastside of Vancouver. *Resources for Feminist Research, 33*(1–2), 19–43.

Brennan, M. A. (2007). Placing volunteers at the centre of community development. *International Journal of Volunteer Administration, 24*(4), 5–13.

Crotty, M. (2003). *The foundations of social research: Meaning and perspective in the research process.* London, UK: Sage Publications.

Farganis, J. (2004). *Readings in social theory: The classic tradition to post-modernism* (4th ed.). New York: McGraw-Hill.

Fay, J. (2003). You can't check your power and privilege at the door: Some lessons in post-modern social justice coalition-building. *Canadian Review of Social Policy, 52*, 139–143.

Fraser, N. (1997). *Justice interruptus: Critical reflections on the "postsocialist" condition.* New York: Routledge.

Freire, P. (2003). *Pedagogy of the oppressed: 30th anniversary edition with an introduction by Donaldo Macedo.* New York: Continuum International Publishing Group.

Gordon, C. (Ed.). (1980). *Power/Knowledge: Selected Interviews and Other Writings 1972–1977 by Michel Foucault.* New York: Pantheon Books.

Grabb, E. G. (2007). *Theories of social inequality* (5th ed.). Toronto: Nelson/Thomson Canada.

Green, L., George, M., Daniel, M., Frankish, C., Herbert, C., Bowie, W., et al. (1995). *Study of participatory research in health promotion: Review and recommendations for the development of participatory research in health promotion in Canada.* Vancouver: Institute of Health Promotion Research, University of British Columbia, and the BC Consortium for Health Promotion Research.

Guba, E., & Lincoln, Y. (2005). Paradigmatic controversies, contradictions, and emerging confluences. In N. Denzin & Y. Lincoln (Eds.), *The Sage handbook of qualitative research* (3rd ed.) (pp. 191–215). Thousand Oaks, CA: Sage Publications.

Hatry, H., van Houten, T., Plantz, M., & Greenway, M. T. (1996). *Measuring program outcomes: A practical approach*. Alexandria, VA: United Way of America.

Israel, B., Schulz, A., Parker, E., & Becker, A. (1998). Review of community-based research: Assessing partnership approaches to improve public health. *Annual Review of Public Health, 19*, 173–202.

Kamberelis, G., & Dimitriadis, G. (2005). Focus groups: Strategic articulations of pedagogy, politics, and inquiry. In N. Denzin & Y. Lincoln (Eds.), *The Sage handbook of qualitative research* (3rd ed.) (pp. 887–907). Thousand Oaks, CA: Sage Publications.

Kincheloe, J., & McLaren, P. (2005). *Rethinking critical theory and qualitative research*. In N. Denzin & Y. Lincoln (Eds.), The Sage handbook of qualitative research (3rd ed.) (pp. 303–342). Thousand Oaks, CA: Sage Publications.

Kirby, S., Greaves, L., & Reid, C. (2006). *Experience, research, social change: Methods beyond the mainstream*. Peterborough, ON: Broadview Press.

Krueger, R. (1994). *Focus groups: A practical guide for applied research* (2nd ed.). Thousand Oaks, ca: Sage Publications.

Lincoln, Y., & Guba, E. (1985). *Naturalistic inquiry*. London, UK: Sage Publications.

Macduff, N., & Netting, E. (2010). The importance of being pracademic. *International Journal of Volunteer Administration, 27*(1), 43–47.

Maguire, P. (1987). *Doing participatory research: A feminist approach*. Amherst, MA: Center for International Education, School of Education, University of Massachusetts.

McCubbin, M., Labonte, R., & Dallaire, B. (2001). *Advocacy for healthy public policy as a health promotion technology*. Toronto: Centre for Health Promotion, University of Toronto. http://www.utoronto.ca/.

Minkler, M., Fadem, P., Perry, M., Blum, K., Moore, L., & Rogers, J. (2002). Ethical dilemmas in participatory action research: A case study from the disability community. *Health Education and Behaviour, 29*(1), 14–29.

Minkler, M., & Wallerstein, N. (Eds.). (2003). *Community-based participatory research for health*. San Francisco: Jossey-Bass.

Morgan, D. (1997). *Focus groups as qualitative research* (2nd ed.). Thousand Oaks, CA: Sage Publications.

O'Connor, M., & Williams, K. (1994). *A workshop on theories and methods of participatory research*. Hamilton: McMaster Research Centre for the Promotion of Women's Health. Workshop held 25 January 1994.

Rahman, M. A. (2008). Some trends in the praxis of participatory action research. In P. Reason & H. Bradbury (Eds.), *The Sage handbook of action research: Participative inquiry and practice* (pp. 49–62). Thousand Oaks, CA: Sage Publications.

Reason, P., & Bradbury, H. (2008). Introduction. In P. Reason & H. Bradbury (Eds.), *The Sage handbook of action research: Participative inquiry and practice* (pp. 1–10). Thousand Oaks, CA: Sage Publications.

Reid, C. (2004a). Advancing women's social justice agendas: A feminist action research framework. *International Journal of Qualitative Methods, 3*(1). http://www.ualberta.ca/.

Reid, C. (2004b). *The wounds of exclusion: Poverty, women's health, and social justice.* Edmonton: Qualitative Institute Press.

Reitsma-Street, M., & Brown, L. (2004). Community action research. In W. Carroll (Ed.), *Critical strategies for social research* (pp. 303–319). Toronto: Canadian Scholars' Press.

Rektor, L. (2002). *Advocacy—the sound of citizens' voices: A position paper from the Advocacy Working Group.* Ottawa: Government of Canada, Voluntary Sector Initiative Secretariat.

Salamon, L., & Lessans Geller, S. (2008). *Non-profit America: A force for democracy* (communique no. 9). Baltimore: Center for Civil Society Studies, Institute for Public Policy, Johns Hopkins University. http://www.ccss.jhu.edu/.

Sayer, A. (1992). *Method in social science: A realist approach* (2nd ed.). London, UK: Routledge.

Stienstra, D. (2003). Listen, really listen, to us: Consultation, disabled people, and governments in Canada. In D. Stienstra & A. Wight-Felske (Eds.), *Making equality: History of advocacy and persons with disabilities in Canada* (pp. 33–47). Concord, ON: Captus Press.

Wallerstein, N. (1999). Power between evaluator and community: Research relationships within New Mexico's healthier communities. *Social Science and Medicine, 49,* 39–53.

Wallerstein, N., & Duran, B. (2003). The conceptual, historical, and practice roots of community based participatory research and related participatory traditions. In M. Minkler & N. Wallerstein (Eds.), *Community-based participatory research for health* (pp. 27–52). San Francisco: Jossey-Bass.

Wight-Felske, A. (2003). History of advocacy tool kit. In D. Stienstra & A. Wight-Felske (Eds.), *Making equality: History of advocacy and persons with disabilities in Canada* (pp. 321–338). Concord, ON: Captus Press.

A PROVOCATIVE PROPOSITION
LINKING RESEARCH, EDUCATION, AND ACTION IN SASKATOON'S CORE NEIGHBOURHOODS

Mitch Diamantopoulos and Len Usiskin

Introduction: Learning the Craft of Community-Based Research Through Case Study Learning

Law students routinely study legal precedents and casebooks to prepare for their future careers. Case study helps them to think logically, develop argumentative skills, and build courtroom intelligence. Similarly, aspiring social scientists and journalists can benefit from studying real-world, community-based research projects. These cases can bring research alive, sharpening critical thinking skills and investigative discipline and better preparing us for the field. Based on a study of community action in Saskatoon's core neighbourhoods, the following provides one such window onto the world—and the democratic craft—of community-based research.

Focus of the Study: The Social Economy Response to the Crisis in Saskatoon's Core Neighbourhoods

In 2006, a research team[1] interviewed about a dozen civil society leaders to explore Saskatoon's "social economy." In the wake of the twin failures

---

1 The team included Mitch Diamantopoulos, Robert Dobrohoczki, and Lori Blondeau, then doctoral students at the Centre for the Study of Cooperatives...

of the market and the welfare state to solve globalization-era economic and social problems, community-based organizations were increasingly tackling problems ranging from unemployment to crime and substandard housing. In the United States, this community-based sector of social enterprises is known as the "third sector" (Rifkin, 1995). In Europe and Quebec, it is known as the "social economy" (Quarter, 1992).

Social economy organizations often emerge to meet needs neglected by the state and private sector, frequently in disadvantaged neighbourhoods—where those needs are most pressing. Taking community action, on a democratic and not-for-profit basis, to provide varied goods and services (Chantier de l'économie sociale, 2010; Quarter, 1992), this sector includes cooperatives, service clubs, charities, and community service organizations. Active in a wide range of activities—from retail, housing, and insurance to childcare, training, and cultural production—these community-based ventures create jobs, provide opportunities for civic participation and democratic skill development, and build social cohesion (Fairbairn & Russell, 2004).

In the postwar era, the needs of Saskatoon's core neighbourhoods became increasingly urgent. By 2006, Saskatoon was a city deeply divided by economic, quality-of-life, and health disparities. Core families were two and a half times more likely than the average Saskatoon family to subsist on less than $20,000 a year by 2002. Fully two-fifths of core families were under that income ceiling. Particularly hard hit were Aboriginal people, who were more likely to be poor, excluded from the labour market, and living in the core. While the average Saskatoon family income had risen to $62,451, the average Aboriginal family income in the core neighbourhoods was $16,497 (City of Saskatoon, 2003). By 2005, infant mortality in

· · · · · · · · ·

...at the University of Saskatchewan. The team worked under the supervision of Dr. Isobel Findlay, Diamantopoulos's co-author for the original study (Diamantopoulos & Findlay, 2007) and university co-director of social economy research at the Community-University Institute for Social Research (CUISR). The team also benefited from the advice and guidance of CUISR's community co-director, executive director of Quint Development Corporation, and co-author of this chapter, Len Usiskin. The study was part of a five-year research initiative on the social economy of Saskatchewan, Manitoba, and Northern Ontario; it was headed by Dr. Lou Hammond Ketilson at the Centre for the Study of Cooperatives, in partnership with other Canadian universities and community and cooperative organizations. Funding was provided by the Social Sciences and Humanities Research Council of Canada (SSHRC).

the core was five times higher than the city-wide incidence. Suicide rates in these poor neighbourhoods were fifteen times higher than the affluent neighbourhood average (Lemstra & Neudorf, 2006, p. 5). Over the past two decades in particular, residents increasingly turned to collective enterprise and community economic development strategies to address the crisis in the core (Diamantopoulos & Findlay, 2007).

Community Entry: Engaging a Vulnerable Community in Transition

We thus entered a divided city—the poorest neighbourhoods of which were mired in crisis—to investigate these community-based efforts. Rather than assemble case studies of discrete organizations, we focused on the "big picture": that is, how the community movement that made up the city's social economy reflected and responded to the city's unequal development. Literature on the core, on problems of inner-city redevelopment, and on the social economy were all reviewed to prepare for interviews, build up a more coherent understanding, and better contextualize the case. The research design was a messy, dialogic process that involved constant matching of theoretical concepts, empirical findings, and perspectives from the field.

Our interview sample was designed to reach into varied subfields so that we could triangulate the city's social economy from various perspectives and build stronger case reliability. In keeping with the emergent design, we also used the "snowball method," asking our informants for referrals to other potential sources. These referrals provided important insights into how organizations were connected. Our modest sample reflected the exploratory intent of the study and the inevitable limitations imposed on real-world projects by busy community partners and limited time, staffing, and budgets.

Contextual Research: A Divided Sector in a Divided Community

The decline of Saskatoon's core neighbourhoods is often treated as some kind of natural catastrophe: a condition without historical or structural underpinnings, a natural and taken-for-granted state of affairs that—like the weather—is simply beyond the reach of policy intervention. In fact, of course, core neighbourhood decline reflected decades of active policy and spending commitments to suburban development. The core was systematically "hollowed out" by investment flight—from postwar suburban shopping mall and residential developments to the big-box pad and gated community developments of the 1980s. In each wave of suburban and exurban sprawl, public infrastructure investments in roads, sewer lines, schools, and recreation centres necessarily paved the way for, and picked up after, private investment.

This expansionist pressure, backed by powerful developers and politically vocal suburbanites, depleted City Hall's resources for inner-city needs. The negative effect of an overextended City Hall on the core was further compounded by the retreat of the federal and provincial governments from social policy intervention in the 1980s. Our study thus situated the grassroots responses of these older, central neighbourhoods—including the organization of a community economic development alliance in 1996—in this formative historical context.

Although researchers often prefer clearly bounded and manageable projects, such as organizational case studies, we soon learned that such pragmatism can not only yield chaotic conceptions that miss the "big picture" but also breed grassroots resentment. In contrast, situating the social economy resurgence within the context of the urban decay cycle's impact on the core eased community entry. We were not embarking on a search for easy-to-manage problems and a better Band-Aid. Rather than approach the crisis in the core as a grab bag of isolated problems that required discrete solutions, we approached issues of employment, poverty, housing, and crime as reflections of the suburban development model. Deregulated market forces were the "first mover" in neighbourhood divestment and decline. As a result of this market-led flight of investment, jobs, services, and opportunities to the outer city, the structure of need—and the centre of social economy innovation—shifted to the city's older core neighbourhoods. Interviews were guided by these findings.

By understanding how the structure of opportunity in these neighbourhoods had collapsed, we could better account for the historical structuring of neighbourhood inequity. Clearly, the concentration of disadvantaged populations in these neighbourhoods was not incidental, unimportant, or, worse yet, the cause of neighbourhood decline—as some anti-poor and racist ideological currents would suggest, blaming the victims for their own misfortunes (Swanson, 2001). Restoring this contextual understanding is of some importance in an age when forty-nine percent of Canadians believe that, "if poor people really want to work, they can always find a job." Twenty-eight percent believe that the poor "usually have lower moral values" and are lazy (Proudfoot, 2011). To overlook the structural roots of inner-city poverty is an ideologically and politically consequential sin of omission, easily recognized by veteran, worldly wise community workers.

We thus viewed this context of neighbourhood divestment, and the development of social economy organization, as crucial to our investigation and to providing more solid foundations for future research. For example, while population health research has shone light on the unequal distribution of good health across Saskatoon neighbourhoods (Lemstra & Neudorf, 2006), a political economy of unequal neighbourhood development further deepens our understanding of urban decay cycles as socio-historical determinants of health. It also highlights the ultimate significance of community economic development mobilizations and the creation of New Social Economy organizations.

The alternative, a naive objectivism that merely treats the historical concentration of predominantly poor and non-white populations in distressed neighbourhoods as somehow natural, functional, or inevitable, reifies a problematic socio-historical reality, renders these processes invisible and unproblematic, stigmatizes those neighbourhoods as "troubled," and invites us to blame the victims—the residents. In fact, the flight of investment, services, and affluent families to the suburbs is essential to an understanding of the social conditions—and the rise of a New Social Economy—in these neighbourhoods.

The Tonto Principle: "What Do You Mean 'We,' White Man?"[2]

Although the social economy organizations that we selected for interviews had much in common, intersectoral affiliation was often weak. Competition for funds or leadership often strained relationships. A high degree of institutional segregation and cross-cultural distrust reflected Saskatoon's colonial legacy.[3] We thus entered a complex, fluid, and often tense cultural context. Sector-specific attachments, discourses, and action frames fractured a sense of shared interest and identity. The traditional, state-centred

- - - - - - - - -

2 We are indebted to Bill Livant for this pithy formulation.
3 The failure of the Core Neighbourhood Development Council (CNDC), an effort to bring together Aboriginal and (formally) non-racialized community organizations in 2001, is illustrative. The Central Urban Métis Federation and the Saskatoon...

service delivery focus of the NGO, the philanthropic focus of foundations and service clubs, the emerging cooperatives' focus on new community needs, and the established cooperatives' focus on retaining and growing market share all drove social economy organizations apart. Front-line social service providers felt little affinity with service club fraternities, and cooperators felt little common cause with trade unionists.

We also found a lack of "bridging social capital"—networks that connect different social groups and thus build their social leverage (Bezanson, 2006). With no unifying organization, these diverse interests remained distant from each other despite their geographic proximity. Similarly, there were difficulties in "frame bridging"—conceptions that build a common point of view among groups that might share interests and values (Carroll & Ratner, 1996). Despite the efforts of the Core Neighbourhood Development Council, which struggled with insufficient resources to involve Aboriginal organizations, other community partners, and government, there was a lack of "social economy consciousness."[4] Groups worked in subfields that were both practically and semantically fragmented.

We learned quickly that the "community" researched was not a community at all. It was a field of conflicting interests, values, and social forces, neither cohesive nor coherent. It was a community in dispersal and dislocation as well as a community in development. The theoretical utility of the term "community" might thus limit and even mislead insofar as it suggests a false unity. First Nations and Métis strove for self-government in the urban milieu, creating Aboriginal organizations to serve Aboriginal needs. Core neighbourhood residents formed the Quint Development Corporation to spearhead inner-city redevelopment efforts. In each case, less powerful groups sought autonomy from the dominant "community" to build new urban communities of interest and take more effective social action. To discuss community-based research in Saskatoon often meant asking in which community the

· · · · · · · · · ·

...Tribal Council deemed the CNDC a funding rival and withdrew from the coalition, and the effort was abandoned (Diamantopoulos & Findlay, 2007). Similarly, in 2008 the provincial government played to these divisions, suggesting that funds released by its withdrawal of support from a multi-use development backed by Quint might be made available instead to urban initiatives of the Saskatoon Tribal Council (Wood, 2008).

4 For an account of the "cultural shifts" necessary for the emergence of Quebec's social economy movement, see Neamtan (2004, pp. 26–30).

research was "based" and in which communities' interests that research was conducted. This created treacherous terrain for community-based research.

Yet we also found that the sense of community itself is in a continuous state of contention and development. For example, the Quint Development Corporation, and an extended social movement family that shared its Community Economic Development (CED) agenda, largely came to define the city's New Social Economy after 1996. The study thus focused on "green shoots" such as Quint, the Saskatchewan Native Theatre Company, CHEP Good Food Inc., the Core Neighbourhood Youth Cooperative, the Core Neighbourhood Development Council, and a major multi-purpose initiative, Station 20 West. This cluster of community economic development action demonstrated emerging needs and a strong sense of belonging to a community movement. We also examined more established cases, such as the Saskatoon branch of the provincial Abilities Council and Saskatoon (now Affinity) Credit Union, rooted in previous historical waves, and contexts, of social action. Finally, we considered the emerging forms of cooperation and conflict between the Old and New Social Economies. Conceptualizing the overlaps and tensions among these distinct social movement families provided sharper insights into the diversity and dynamism of Saskatoon's social economy. It also helped to guide the selection of respondents.

Far from passively reflecting an inert, objective reality, the processes of conversation, publication (Diamantopoulos & Findlay, 2007), and presentation (Warren, 2010) that comprised this study were active interventions in the reflections of an unfolding social movement for economic action, in the subjects' thinking about their work, in their relationships with other organizations, and in how they conceived the possibilities for joint action within that field. As we carried out our interviews, it became clear that the research task was thus also a task of education and action: providing a mechanism for "pooling" community experience and strategic reflection as well as merely applying documentary insight and expertise. As our understanding of the need for dialogue grew, our commitment to sharing relevant and accessible findings also grew. The study became a progressively more purposeful project in community development.

Building Democratic Social Capital: Agitate, Provoke, Propose

We discovered that systemic misunderstandings about the nature and origins of core neighbourhood problems contributed significantly to those problems. This crisis was also a crisis in our thinking about the core. The preponderance of news coverage of crime in the core on the one hand and boosterism for suburban expansion on the other helped to feed a view of core residents as demonized "others"—drug-addicted, welfare-dependent slackers—who posed a moral and criminal threat to the larger community. This tendency to mythologize reduced public issues to matters of personal choice and responsibility, fed the stigma of core "otherness," and encouraged politicians to exploit these fears. Predictably, this narrative of the core—as populated by unrealistically dysfunctional and dangerous people—increased distrust of prying journalists and social scientists.

But resistance to "outsider" investigation had three other wellsprings. First, there was a perception that public funds were too frequently steered into studies that did not improve core neighbourhood conditions. Research was viewed as the state's excuse for not taking real action. Many felt that the core neighbourhoods, or Aboriginal people, "had been researched to death." Second, front-line community leaders saw "prudent" study limitations at best as expedient for time-strapped academics or at worst as political evasion of "root causes" and possible controversy. Third, state-funded organizations routinely endure probing evaluations. "Administrative research" (Smythe & Van Dinh, 1983) focuses on organizational efficiency, but this definition of the problem and adoption of an evaluative method can provide a pretext for funding cuts. Shining yet another bright light on underresourced, overextended, and vulnerable groups can thus generate defensiveness and resistance. Since administrative evaluation also tends to "let the state off the hook" for misguided policies, it can feed generalized hostility toward researchers, particularly academics. We took extra pains to build trust and ease these anxieties.

As Stoecker (1999, p. 842) argues, participatory research requires a two-way dialogue, combining the researcher's theoretical knowledge with the insider's first-hand knowledge of the milieu. Instead of adopting an authoritarian pose of scientific expertise and a "culture of lecture," or a pandering pose of uncritical appreciation and one-way listening, Kleidman (2009, p. 353) recommends a "culture of agitational conversation" to bridge grassroots experience and academic expertise:

> *In working with activists, a culture of conversation avoids both extremes—those of imposing our views, and those of deferring completely to those of our partners. We should develop relationships that*

allow us to take an active role toward community partners, using social movement theory and research to provide a critical analysis of their work. However, we should present our ideas as hypotheses and suggestions, and seek and listen to responses that can also be challenging.

The International Institute for Sustainable Development (2007, p. 3) suggests that collaborative inquiry be driven by "provocative propositions," realistic dreams that "challenge [sources] to move ahead by understanding and building on their current achievements." In contrast to the faux objectivity of documenting an ahistorical and de-subjectivized phantom community—populated by facts, figures, and problems alone—this approach highlights the importance of our interpretive understanding of actors' identities, aspirations, and strategic conceptions. Where administrative research isolates, evaluates, and depoliticizes social economy organizations, critical community-based research can thus reconnect, repoliticize, and revitalize community movements for economic action—and demonstrate the legitimacy of their work in the broader community and corridors of power. This amounts to the difference between a managerial approach to proprietary knowledge contracted by the state, which compartmentalizes research into organizational studies, and a more democratic and developmental approach, which publicizes research that is wider in its scope.

Recognizing the reality and roots of respondents' resistance, interviewers took care to establish the broad scope of the investigation and to set a democratic tone with the methods of agitational conversations and provocative propositions. Demonstrating respect for their important work and perspective as community leaders, this distanced the investigation from their experience with evaluative research and helped to build trust, goodwill, and openness or "social capital" (Coleman, 1988).

Similarly, since "social economy" was a foreign and alienating term to many, more familiar terms communicated our eagerness to learn from grassroots expertise. We were not aloof scholars out to verify preconceived hypotheses by "testing" them. Rather than undertake evaluative research based on a client relationship and an expert pose, we approached interviews in a democratic spirit of joint investigation.

Building credibility, trust, and rapport also depended on asking the "right questions." By building on prior community learning about the need for broader investigation, we demonstrated that we were listening, were learning, had done our "homework," and were taking community expertise seriously. Rather than counterposing community feeling and scientific detachment, we conducted semi-structured interviews that gave wide latitude to our partners' experiences, interests, and priorities.

Sustaining Democratic Partnerships: The Role of Institutional Intermediaries

As a focus of market and state failures, central city neighbourhoods increasingly drove resurgent social economy action in the 1990s. However, community divisions and conflicts made gaining access and goodwill difficult. In a distressed community, rife with distrust, fragile funding, and strained interorganizational relationships, anxieties can run high. Researchers can help to identify common ground, clarify misunderstandings, and find new ways to work better together, but they can also become lightning rods.

CUISR was an important institutional intermediary. It had not only built up important local knowledge over the years but also facilitated community-university social capital. Through this network of scholar-activist relationships, CUISR bridged "structural holes" (Burt, 2004) separating diverse networks of interaction, relationship, and information flow, enabling previously divided communities to create a safe haven to exchange views and build relationships. These dialogues on community development also created important bridging frames between Aboriginals and non-Aboriginals, trade unionists and cooperators, and NGO staff and church staff. Over time, these joint involvements in community-based research cultivated intersectoral trust, mutual understanding, and a sense of shared mission and responsibility. In this culture of community-based problem solving, community-based research was valued.

These conversations, joint projects, and occasional publications also bridged the structural hole that segregates academics and community leaders. They helped to shape a common local language and practice for community-based research, an important social achievement. CUISR helped to transcend the twin pitfalls of "verbalism," theory ungrounded in practical action, and "actionism," practice uninformed by theoretical reflection (Freire, 1973). It became an active site for research that closed "the gap between studying a problem and identifying hands on solutions" (Silka, 2003, p. 61). Providing grassroots access to university resources was an important conduit for the formation of "leveraging social capital"

(Bezanson, 2006) that would mobilize community-university relationships for community development.

In this study, another important intermediary was a Saskatoon city paper, *Planet S* magazine. The lead author took a leave from the paper to take on this study, benefiting from contacts, contextual understanding, and interview skills built up through his work at *Planet S* and the trust and street credibility conferred by association. *Planet S* was both structured as part of the New Social Economy (as a worker cooperative) and culturally rooted in the community movements (as an alternative city paper that had spilled considerable ink on core revitalization efforts).

Walking the Walk: The Role of Critical Reflexivity

Of course, the occupational hazards of the community-university partnership remain, and so does the need for reflexive vigilance. Structural tensions based on conflicting interests, incentives, and needs persist. Paradoxically, we heard complaints about being both "researched to death" and "neglected" by a self-serving academic establishment—an index of the difficulties of community-based research.

Several factors undermined meaningful engagement. First, many sources worked for understaffed organizations and were too busy to meet with us or simply did not see the benefits of doing so. Second, some resented a scholarly class whom they viewed as tourists—a knowledge elite jetting off to international conferences while ignoring urgent social problems in their own backyard. Winning trust and creating a space for frank dialogue were key challenges.

Establishing trust and rapport takes time, which is often in scarce supply for academics and journalists alike. They must juggle multiple projects and move through serial commitments, even though the social problems that they study remain unsolved. The churning of graduate students creates greater difficulties for academic investigations. This revolving door of fleeting academic commitments is frustrating, both to sources and to investigators, and can deplete hard-won social capital. The small scale of this study, conducted over a few weeks, was symptomatic. Against these institutionalized academic pressures, intermediaries such as CUISR help to build effective partnerships.

Incentive structures also pull academics away from the community. Pursuit of tenure and promotion encourage faculty to orient their work to national and international specialists rather than local needs or popular audiences. Local engagement thus cuts against the grain of well-established career paths.

Conversely, in our eagerness to build strong community partnerships, academic researchers can too easily romanticize maximizing partner participation. In reality, respect also needs to mean respecting limitations that prevent more full participation. In this study, interviews routinely ran from two to four hours with very busy sources. A perfectly participatory research design would have to build on a narrower, less representative base and consume extra hours for already overstretched community partners.

Different time horizons further exacerbate "town and gown" tensions. While community leaders often fight for their organizations' day-to-day survival, academics routinely spend years on a single book. In contrast to criticisms of event-driven, "drive-by journalism," data collection for this study was completed in the summer of 2006, but the limited-circulation report was not issued until August 2008 (Diamantopoulos & Findlay, 2007). As a result of publication delays and the lead author's relocation to take a job out of town, it was April 2010 before a public presentation was staged and the findings reported to the public (Warren, 2010). The reasons for academic publication delays, like the need for journalistic speed and concision, seldom comfort grassroots actors.

While traditional academic scholarship has clear guidelines for reliable and valid inquiry, community-based research implies a dual accountability:

1. to provide an account that conforms to the rigorous rules of logic, evidence, and ethical inquiry; and
2. to contribute meaningfully to community development by developing an engaged and informed citizenry.

Making Research Count, Making Research Public

The final study report was written with a democratic intent that gave voice to front-line social actors. Publicly accessible prose informed public debate. Rather than write up findings for academic journal publication, we paid special attention to grassroots examples, anecdotes, and quotations to lend popular realism and resonance to the narrative. This approach highlighted the valid experiences, interpretations, and leadership roles of social economy actors—too frequently marginalized by the cultural hegemony of "official sources" in the news and naive objectivism in the academy.

Community-university partnerships often lack the resources, expertise, or inclination to publicize their findings. To this extent, community-based research projects can remain insider affairs. Participants might not recognize the validity of wider public engagement or the role of journalists as popular interlocutors. These closed loops highlight the need to deepen democracy by more effectively making knowledge public. This implies that community-based research become an educational site for community partners, scholars, and journalists. Community-based research becomes a place where community partners build analysis and voice, journalists build stronger sociological sensitivity and connection to the community and the academy, and academics ground themselves in the community and enhance their ability to address a democratic public and its press better. Community-based research might thus be understood less as a paternalistic practice of "community service" and more as a democratic engagement, an opportunity for a "border pedagogy" (Giroux, 1988), where groups learn new skills from each other.

Community-based research provides a natural place to meet for the deeply segregated research traditions and interpretive communities of social science, social work, and journalism. Arguably, community-based research is as incomplete without the skills of popular engagement codified in the democratic craft of journalism as it would be without the intellectual rigour provided by sociology and scientific method. We are far from bridging this divide, but the importance of doing so seems to be undeniable if we are to put the public back into our research, education, and action on public issues.

Conclusion: It's a Tough Job

Five key ethical and methodological themes emerge from this case. First, community-based research must make sense and matter to the community, reflecting its needs and priorities. Researchers must thus build a contextual understanding of that community's history, divisions, and problems that most cry out for solutions.

Second, community-based research is more than research geographically situated in the community; it is part of a process of community development as surely as academic inquiry is part of developing the scholarly field. This implies dual accountabilities and dual skill sets. Community engagement requires patience, persistence, and goodwill to weather the institutionalized conflicts and cultural differences that routinely dog community-university partnerships and journalist-source relations. Finding mutual benefits in community research, whether it's an academic developing a research program or a reporter building a "beat," therefore requires building social capital, including stocks of trust, open lines of communication, and norms of mutuality (Coleman, 1988).

Third, institutional intermediaries can help to build research partnerships. Without community-university research institutes and alliances (or investigative media), such efforts would be more difficult, initiatives more isolated, and findings less likely to cohere in a body of local knowledge. Building cumulative local understanding without these intermediaries would be as difficult as building a scholarly literature without academic journals, associations, and conferences.

Fourth, community-based research requires critically reflexive diligence. Scholars must be ever conscious of the extent to which their work takes place in a contradictory social location; it is embedded in a structure of academic incentives that drives them down competitive and individualistic publication paths, into national and international peer relations, and away from a focus on local communities' needs, popular publication, and citizen review. Academic knowledge production structures fields of inquiry around fragmented disciplinary specialties. This compartmentalizes knowledge around conceptual rather than social problems (Fairbairn & Fulton, 2000), leading to a focus on the structure of the scholarly field, its gaps, its depths, and its errors of fact and interpretation. Navigating the contradictory pull of theory and practice without being pulled into an anti-intellectual community boosterism on the one hand or a hyper-abstracted academic specialization on the other is no small struggle.

And fifth, community-based research needs to look beyond the particular interests of discrete partner organizations to a third "partner," too often invisible, in community-based research: the public. Efforts to truly democratize knowledge and empower communities must be more than an effective interprofessional collaboration between the managers of community-based organizations and academic professionals. Instead, research, education, and action must also be considered part of the democratic project to help local publics solve social problems, enfranchising citizens with the evidence and conceptual understanding that they need to make sound choices. In this process, public intellectuals drawn from the community, the academy, and the press all have important roles to play.

References

Bezanson, K. (2006). Gendered relations: Gender and the limits of social capital. *Canadian Review of Sociology and Anthropology, 43*(4), 427–444.

Burt, R. S. (2004). The social capital of structural holes. In M. F. Guillen, R. Collins, P. England, & M. Meyer (Eds.), *The new economic sociology: Developments in an emerging field* (pp. 148–192). New York: Russell Sage Foundation.

Carroll, W. K., & Ratner, R. S. (1996). Master frames and counter-hegemony: Political sensibilities in contemporary social movements. *Canadian Review of Sociology/ Revue canadienne de sociologie, 33*, 407–435.

Chantier de l'économie sociale. (2010). *The social economy in Québec.* http://www.chantier.qc.ca/.

City of Saskatoon. (2003). *Neighbourhood profiles* (7th ed.). Saskatoon: City of Saskatoon.

Coleman, J. S. (1988). Social capital in the creation of human capital. *American Journal of Sociology, 94*, S95–S120.

Diamantopoulos, M., & Findlay, I. (2007). *Growing pains: Social enterprise in Saskatoon's core neighbourhoods: A case study.* Saskatoon: Centre for the Study of Cooperatives and CUISR. http://www.usask.ca/cuisr/.

Fairbairn, B., & Fulton, M. E. (2000). *Interdisciplinarity and the transformation of the university.* Saskatoon: Centre for the Study of Cooperatives.

Fairbairn, B., & Russell, N. (2004). *Co-operative membership and globalization: New directions in research and practice.* Saskatoon: Centre for the Study of Cooperatives.

Freire, P. (1973). *Pedagogy of the oppressed.* New York: Continuum.

Giroux, H. (1988). Critical theory and the politics of culture and voice: Rethinking the discourse of educational research. In R. R. Sherman & R. B. Webb (Eds.), *Qualitative research in education: Focus and methods* (pp. 190–210). New York: Falmer.

International Institute for Sustainable Development. (2007). *From problems to strength: Appreciative inquiry and community development.* http://www.isd.org/ai/default.htm.

Kleidman, R. (2009). Engaged social movement scholarship. In V. Jeffries (Ed.), *Handbook of public sociology* (pp. 341–356). Lanham, MD: Rowman and Littlefield.

Lemstra, M., & Neudorf, C. (2006). *Health disparity in Saskatoon: Analysis to intervention.* http://www.uphn.ca/.

Neamtan, N. (2004). The political imperative: Civil society and the politics of empowerment. *Making Waves, 5*(1), 26–30.

Proudfoot, S. (2011, March 1). Poverty myths rampant: Report. *Leader-Post*, p. A7.

Quarter, J. (1992). *Canada's social economy*. Toronto: Lorimer.

Rifkin, J. (1995). *The end of work: The decline of the global labor force and the dawn of the post-market era*. New York: Putnam.

Silka, L. (2003). Community repositories of knowledge: A tool to make sure research pays off for university partners. *Connections*, (Spring), 61–64.

Smythe, D. W., & Van Dinh, T. (1983). On critical and administrative research: A new critical analysis. *Journal of Communication, 33*(3), 117–127.

Stoecker, R. (1999). Are academics irrelevant? Roles for scholars in participatory research. *American Behavioral Scientist, 42*, 840–854.

Swanson, J. (2001). *Poor-bashing: The politics of exclusion*. Toronto: Between the Lines.

Warren, J. (2010, April 14). Social groups doing work to resuscitate inner city: Researcher. *Star-Phoenix*, p. A11.

Wood, J. (2008, April 18). Saskatoon Tribal Council backs Station 20 funding cut. *Star-Phoenix*, p. A6.

IMPACT OF COMMUNITY-BASED RESEARCH

THE CUISR-SRIC COLLABORATION
TOWARD COMMUNITY-BASED ACTION RESEARCH?

Lori Ebbesen, Janice Victor, Louise Clarke, and Nicola Chopin

Introduction

This chapter is about a multi-year collaboration between the Community-University Institute for Social Research (CUISR) and the Saskatoon Regional Intersectoral Committee (SRIC). The two have collaborated since the fall of 2007 on the development and implementation of a multi-phase, multi-dimensional action-learning evaluation framework examining the processes and impacts of SRIC. This sustained program of evaluation has conformed, to varying degrees, to the basic tenets of CBR, grounded in alignment on these basic tenets. It has been our experience and premise that alignment is a multi-layered construct that includes alignment *beyond* the collaboration, *between* the partners, and *within* each partner. Furthermore, as it has been for CUISR and SRIC, building on and maximizing the alignment between partners on the tenets of CBR are central to a fulfilling and successful collaboration.

Alignment and misalignment have shaped the CUISR-SRIC journey of collaboration, a journey that evolved from a more traditional client-evaluator model (Bryk, 1983) toward one of participatory evaluation (Cousins & Whitmore, 1998), more congruent with CBR. Critical to this shift toward an action-learning paradigm was a parallel and complementary path toward greater alignment between the partners on the tenets of CBR—especially action orientation—and attention to areas of misalignment. We moved to a more dialogic, flexible partnership as the scope of SRIC's action—and, therefore, the participatory evaluation research—expanded accordingly.

In the sections that follow, we describe the separate and interdependent journeys of CUISR and SRIC in this action-learning collaboration and reflect on the lessons gleaned throughout our journey from "practical" toward "transformative" participatory evaluation (Cousins & Whitmore, 1998). Our aim is to assess this collaboration as it relates to and exemplifies the underpinnings of CBR. The version of CBR that CUISR has adopted is participatory action research, the main tenets of which are problem identification by the community, dialogic research planning, participation as process, co-production of knowledge, capacity building, knowledge translation, and knowledge for change or advocacy (Elliott, 2011).

Background/Context

CUISR's origin is in the introduction to this volume, and only those aspects of note to the SRIC partnership are highlighted here.

First, the people from the community and university who formed CUISR were committed to making the relationship as equal as possible while residing in—and being accountable to—an academic setting. The structure is based on co-directors, management, and advisory boards that have fifty percent community and fifty percent university representation. Core funding for the institute was provided by both the University of Saskatchewan and community partners—the City of Saskatoon, Saskatoon Health Region, and United Way. With the first Community-University Research Alliance (CURA) grant (2001–07), funds were available each year for community-initiated projects and research "sabbaticals" for community partners. The second major CURA (2006–12), this time to study the social economy in Saskatchewan, had an even greater focus on community engagement in the entire research process.

Second, CUISR has always been place based, initially with a focus just on Saskatoon. In 2007, as part of the university's Integrated Planning process, CUISR developed a strategic plan with five strategic directions that expanded the sense of place from the local through provincial to national and international levels. One of the strategic directions, "Sustainable Saskatoon," was a continuation of quality-of-life work and related initiatives on poverty reduction, especially housing and homelessness, while understanding the broad context within which they occur.

Third, CUISR has addressed the need to diversify funding sources because core funding from the university was winding down. It began to receive requests to do research from various community-based organizations, one of the first being SRIC.

SRIC is one of ten regional intersectoral committees established in 1995, linked to the Government of Saskatchewan's Human Services Integration Forum and mandated to work in partnership to coordinate linkages that shape and influence program, policy, funding, and resource deployment to meet the diverse needs and interests of the City of Saskatoon and twenty-five surrounding rural municipalities. It has gradually grown to twenty-six human service leader members from a wide range of agencies, including provincial ministries, federal departments, the City of Saskatoon, Saskatoon Tribal Council Urban Services Inc., the Central Urban Métis Federation Inc., the Saskatoon Police Service, the Saskatoon Health Region, boards of education, two post-secondary institutions, and several NGOs such as the United Way. CUISR is one of the two University of Saskatchewan representatives and joined in 2005. SRIC has co-chairs, one from the provincial government, one from a local agency or NGO, with staggered two-year terms. It meets about five times a year, and decision making is by consensus. Operations, including the salary of a full-time coordinator and a part-time administrative assistant, are funded by the Human Services Integration Forum and a levy on all members based on their ability to contribute financially. The coordinator and/or members participate on a number of steering or advisory committees in the region, resulting in a dense network of service providers.

Our Journey

SRIC and CUISR have experienced two relatively distinct phases of building our research relationship: an early one between 2007 and 2008, which was similar to a client-consultant relationship (Bryk, 1983), and a transitional period in 2009, during which several important events occurred, shifting the relationship toward one more in tune with a CBR partnership.

Early Phase of the Partnership

In the fall of 2007, the SRIC coordinator approached CUISR about evaluating some project work that it was already doing and implementing its new strategic plan. The coordinator met several times with CUISR's research team to discuss the work required. This team was comprised of the research coordinator (staff), one or two faculty advisers, and, at some points, one or two graduate student interns. Key elements of the first work plan were a review of a grant program administered by SRIC, interviews with selected SRIC members to obtain their views of how its new mission and purpose could be enacted and which measures would indicate success, and a three-year cycle of evaluation activities. When CUISR presented its work at the SRIC meeting in May 2008, discussion focused on the proposed evaluation framework, and there was consensus that it had to be less academic and more practical. Meanwhile, SRIC had formally adopted its new strategic plan, intended to expand the committee's focus to acting on, not just discussing, ways to improve service coordination. Over the summer and fall of 2008, CUISR worked on a revised plan for evaluation that incorporated suggested revisions and considered SRIC's intent to develop its action orientation.

Turning Points in the Partnership

In early 2009, three key developments in our journey occurred that catalyzed the CUISR-SRIC collaboration: SRIC approved a revised framework for evaluation, embarked on a new action orientation, and convened an Evaluation Working Group.

The CUISR team, accepting that a clearer framework was needed to overcome SRIC members' resistance and skepticism, developed one that depicted the task of evaluation holistically as three concentric circles, reflecting the committee's various rings of influence, with clear research questions and methods identified for each circle. The inner circle depicted SRIC members and processes; the middle circle comprised the people and organizations to which its members report; and the outer circle included SRIC's geographic region and citizens (see Appendix A for a fuller description of the framework). This creative yet practical framework was effective in receiving the support of SRIC members.

Historically, the committee played a behind-the-scenes role in the community, focusing primarily on discussing ways to improve the coordination of human service delivery. Prompted by a health disparities report published by the Saskatoon Health Region (Lemstra & Neudorf, 2008), SRIC prioritized action on three key policy areas identified in the report: poverty reduction, Aboriginal employment, and housing and homelessness,[1] marking a shift from discussion to direct engagement. SRIC decided to play a supportive role in poverty reduction efforts and the subsequent working group since there was already a community coalition actively trying to develop initiatives. With the Aboriginal employment work, there was no pre-existing community coalition, so SRIC decided to take the lead role in forming a working group. These diverse action leadership roles presented an interesting comparison in terms of evaluation.

Next, SRIC formally convened an Evaluation Working Group (EWG) tasked originally with providing advice on various research issues, guiding the research collaboration, and ensuring that the evaluation is grounded in SRIC practice. It had six members representing government, non-government, and academic perspectives; SRIC co-chairs are ad hoc members welcome to participate as they wish. The group expanded its role and became influential in nurturing a culture of evaluation by regularly presenting evaluation plans and progress and fostering reflexive discussion. An update from the EWG became a standing item on the agenda for each SRIC meeting, an opportunity to inform members and, more importantly, to engage the committee in discussions about the findings and their action implications, thereby generating more collective support and enthusiasm.

Methodologically, the CUISR-SRIC partnership embarked in two directions. First, we launched the RIC Self-Assessment Tool (RSAT) developed by CUISR to measure members' assessments of their internal processes, a key element of the "inner ring" of the evaluation framework. The RSAT, administered online, has forty-three questions measuring synergy and member contributions, leadership, efficiency, and management and requires about ten minutes to complete. In May 2009, the EWG presented the findings and facilitated small-group discussions to elicit reactions and action items from SRIC members that, in turn, became additional sources of data.

Second, the EWG seized the opportunity for evaluation presented by SRIC's varied roles in the working groups. It persuaded SRIC's members to add the new poverty reduction and Aboriginal employment strategies to the evaluation framework and to adopt a forward-looking, learning approach to comparing SRIC's role in supporting the former (likened to "riding the bus") and leading the latter ("driving the bus"). From the fall of

----

1 There were forty-six evidence-based policy options presented in the report. SRIC members initially identified a specific priority, Aboriginal youth employment, later framed more broadly as Aboriginal employment.

2009 until June 2010, the CUISR team carried out participant observation in the working groups and interviewed SRIC members purposely selected based on various levels of engagement with the working groups.

Benefits of the Partnership

To date, the CUISR-SRIC collaboration has been beneficial for both partners. The action research process sparked a higher level of engagement between the partners and among SRIC members; generated new knowledge useful for us together and separately; enhanced research capacity through the integration of new methodologies and perspectives; and expanded each other's knowledge of practical benefits and challenges of community-university partnerships for evaluation. In short, this journey took us toward ever-greater alignment of purpose and relationship values.

Outcomes/Lessons Learned

The CUISR-SRIC journey has taught or reminded us of various aspects integral to community-based collaboration. Reflection on our journey of the past few years revealed several lessons that can guide us in our future work:

- identifying areas of alignment;
- acknowledging areas of misalignment; and
- understanding the role of the EWG.

Reflection also allowed us to assess how closely our work truly exemplifies CBR.

Identifying Areas of Alignment

The CUISR-SRIC collaboration has benefited from several key areas of alignment. As we have discovered, some aspects of alignment go *beyond* the partnerships and are situated within the parent body to which each partner is ultimately accountable (e.g., mandates, missions, priorities). Some aspects of alignment fall *between* the partners and are open to negotiation in traversing a smooth path together (e.g., action orientation, relationship parameters). Still other aspects are *within* partners and speak to internal dynamics (e.g., shared vision, cohesion).

In our case, beyond partner aspects have been more alike than different, providing a strong launching pad for collaboration. Both partners have strong core values driving our work: commitment to the community (or communities) in which we operate, equity, inclusiveness, accountability, and fairness. We both have missions to address social issues, working with or on behalf of marginalized groups, and we are accountable, in part, to the same geographic community. CUISR and SRIC are established and respected entities in Saskatoon and area, known for our commitment to improving social conditions.

Between partners, CUISR and SRIC have brought genuine interest in and commitment to developing a functional and productive relationship. To that end, there has been mutual attention to clarifying roles and responsibilities; devising joint action plans; using varied forms of evidence and subsequently creating a range of knowledge "products"; advancing dialogue about implications of our work; and airing differences. Not all of these aspects are aligned perfectly; however, on most aspects, CUISR and SRIC negotiate ways of acting that are agreeable to each partner.

Between and within partners, this collaboration has proven to be a testing ground of sorts, an opportunity to build respective capacity and, subsequently, to influence each other's ways of working. Through engagement with the working groups, both partners learned about intersectoral initiatives with diverse participants and how best to evaluate SRIC's role and impact in those initiatives. Over the course of our journey, SRIC has come to appreciate the merits of qualitative methods to capture and explain subtle processes of community change.

Within SRIC, evaluating the impact of its own work was a progressive step. The results and discussion of the self-assessment tool directly and indirectly improved its ability to orient new members and communicate effectively, including the use of interactive discussion during meetings (in contrast to straight reporting). Discussing the evaluation framework and the inclusion of the working groups within it increased members' understanding of evaluation and the benefits of the action-learning approach to evaluation in particular.

Within CUISR, the work with SRIC built capacity in two key respects. First, the work is multi-dimensional and somewhat open ended (as opposed to being project specific and time limited), so CUISR has had the opportunity to develop some innovative approaches and methods. Second, as mentioned, SRIC was one of the first groups to pay for CUISR's services (as opposed to the work being covered by research grants), so the institute has increased its capacity to manage this type of relationship in general and

SRIC's requirements in particular as well as to navigate a different administrative system at the university.

Of particular importance in this collaboration, within partners, there has been shared enthusiasm for both innovation and reflexive learning. We have engaged in thoughtful examinations of both processes and outcomes intrinsic to separate and collective work, learning from lessons along the way and adjusting action paths as a result.

Acknowledging Areas of Misalignment

In our sustained CUISR-SRIC partnership, as expected, "rough spots" have arisen from areas of misalignment. Most have been relatively minor and have already been resolved, but some operational and substantive misalignments have required more mutual accommodation.

Operationally, we experienced misalignment between partners related to the timing of our work. The cycles of activities for CUISR and SRIC had some parallels since both tend to accelerate action beginning in September; however, timetable misalignment has been evident in preparing for respective fall activities. For CUISR, setting up internships with students requires earlier planning when it comes to identifying evaluation tasks and related qualifications and skills in order to recruit effectively and establish a viable plan of work. Evaluation planning with SRIC was later than ideal for student recruitment purposes and was resolved, quite simply, by shifting our planning to as early as May for the next year.

Substantively, CUISR's mandate is to do research that is academically rigorous yet adopts the tenets of good CBR, including advocacy. This means that the research is grounded in both the relevant academic literature and the data, which are collected using sound ethical methods. Indeed, the work is of most value to its community partners only when this is achieved. CUISR is also committed to consulting extensively with partners about the research and to training and mentoring students in CBR. Partners such as SRIC with more urgent, practical needs and used to dealing with private consultants might not fully understand these requirements, creating some tensions. And, while there are practical limits to the time and resources to do CBR using research funds, these limits are usually and understandably stricter when a community group is paying for the work. It is understandable, then, that some misalignment arose between the two regarding roles and expectations and the pace of completing the proposed areas of study. Although SRIC members might not have appreciated all the differences between CUISR and a consultant, they did appreciate its ability to contribute to and change the research plans, and they have shown flexibility on timelines and budgets in response to those changes.

Within partner misalignment has been related specifically to advocacy as an integral part of CBR work. SRIC, despite its shift in action orientation toward increasing engagement in community action, is experiencing some lingering dissension over and discomfort with this direction among its members, since many are constrained by their positions as senior government administrators. CUISR, as a proponent of CBR, sees itself as an advocate for policy change based on its research findings, so it is internally aligned. Both partners are centrally involved in the poverty reduction partnership, and there might come a time when there is overt misalignment between the two partners over an agenda of change. This is not necessarily a bad thing; rather, it could be an opportunity for open debate on strongly held views and values.

Understanding the Role of the EWG

The EWG has been pivotal in fostering the momentum necessary to shift our CUISR-SRIC relationship from client-evaluator to partners. It has guided and transformed the research partnership and addressed hurdles along the way, whether they are differing perspectives on substantive issues or process related. The EWG has been instrumental in shaping our journey from "practical" toward "transformative" participatory evaluation—the former characterized by supporting program and organizational decision making, and the latter having a foundation of social justice (Cousins & Whitmore, 1998).

The EWG is a blend of researchers and other community stakeholders with stakes in the foci of the evaluation. By virtue of its makeup and inherent consideration of various perspectives, it ensures research rigour, brings a rich understanding of SRIC expectations, and satisfies minimal definitional requirements to accomplish participatory evaluation (Alkin, 1991; Cousins & Whitmore, 1998). Our initial, pragmatic role, to ground the evaluation in SRIC's program activities, adheres to the purposes of "practical" participatory evaluation (Cousins & Whitmore, 1998; Garaway, 1995). We pushed the limits of a pragmatic function as the EWG role expanded over time. Increasingly, its members deepened their participation, becoming more involved in setting the research agenda, design, and measurement and interpretation of results. The group became the hub of the co-creation of knowledge, engaging SRIC members in iterative dialogues, as with the RSAT, creating additional data and insights, and drawing out implications for practice and policy. The translation function increased as the EWG explicitly devoted attention to building understanding of the research process itself, demystifying it for some and adding to the methodological and practical tool kits of those more experienced. The working group participated in reporting plans and findings to SRIC, which helped to legitimate the research findings to SRIC members, especially in presenting sensitive findings. The EWG became—and remains—a critical intermediary solidifying the CUISR-SRIC partnership, fostering a shared purpose, creating a supportive infrastructure, and sharing responsibility for the evaluation processes and results, all roles closer to the depth and type of engagement typical of "transformative" participatory evaluation (Cousins & Whitmore, 1998). Although we satisfy these components of transformative participatory evaluation, we fall short on our ability to accomplish fully its underlying function of social justice. That said, the SRIC foray into the working groups has the potential to move us farther along this path.

Exemplifying CBR

The version of CBR that CUISR espouses in its work with SRIC and other community partners is participatory action research (Elliott, 2011), the main tenets of which are

- problem identification by the community;
- dialogic research planning;
- participation as process;
- co-production of knowledge;
- capacity building;
- knowledge translation; and
- knowledge for change or advocacy.

To varying degrees, the CUISR-SRIC collaboration has been true to these tenets of participatory CBR. The collaboration that has evolved certainly falls within broad CBR definitions with respect to bringing together community and academic expertise to identify social needs and priorities, advance understanding, and co-create knowledge that makes a constructive difference in the world (Flicker, Savan, Kolenda, & Mildenberger, 2008; Roche, 2008). The EWG has contributed substantially as an intermediary for knowledge translation, successfully advancing this function for and with the partners and seeking ways to be more influential in policy-related discussions. There remain, however, two aspects of our collaboration within the framework of CBR that many will view as controversial: SRIC is not "community" in the commonly understood sense as marginalized, and the intentions and outcomes to date might not be advocacy and transformation.

We argue that, although SRIC members are clearly not a group of marginalized people, they are nonetheless a community in that they interact within a social network (SRIC) formed around a set of common interests (Heller, 1989). Moreover, they have a deep sense of belonging and commitment to the well-being of the people of the geographic community of Saskatoon and region (McMillan & Chavis, 1986). And service providers are often engaged in CBR (Flicker, Savan, Kolenda, & Mildenberger, 2008). Often these service providers work at or close to the front lines, but why should senior-level service providers be automatically excluded as part of CBR? Indeed, SRIC members have great potential to effect the kind of program and even policy changes that are the objects of CBR. The extent to which the CUISR-SRIC participatory evaluation will contribute to actual program and policy change is an open question; however, since the beginning of our partnership, SRIC members have moved a long way from being primarily receivers of information from the community toward being supporters of and advocates for change in that community.

Conclusion

The multi-year partnership between CUISR and SRIC has been rich in experiential learning and successful in advancing a broad scope of innovative work. It has revolved around an inclusive definition of community and multi-layer alignments *beyond*, *between*, and *within* partners. Aspects of alignment and misalignment have also flavoured the collaboration. Our journey together has been catalyzed by a variety of turning points, influential because they have served as opportunities to take stock of and foster greater alignment. We strive to honour and implement more closely the basic tenets of CBR, and we recognize the pivotal intermediary role played by the EWG in this continued journey.

Our case study positions the various layers of alignment as essential ingredients in sustained CBR. CBR is iterative and non-linear; so, too, is alignment. Distinctions between states of alignment and states of misalignment are arbitrary, and, as our CUISR-SRIC journey illustrates, the more likely scenario in a sustained collaboration is an ebb and flow of alignment and misalignment. We entered into this collaboration with a strong sense of alignment beyond partners, starting from a similar commitment to social services that enhance the quality of life for citizens of the Saskatoon region. Early on, we diverged in our alignment between partners in terms of framing a viable evaluation model and plans but converged on those items in a second iteration. We also diverged in our commitment to an action orientation in our respective fields of work, but SRIC members' engagement with the RSAT and the working groups has made possible an action-learning approach to the evaluation.

Despite SRIC's shift in orientation toward engaging directly in community action, there is lingering discomfort with this direction among some of its members, as some express eagerness to move forward to advocate for policy change. This discomfort means that not all members have embraced our action-learning paradigm, and it continues to create some fragility in our alignment between partners. Questions have been raised, in the context of the working groups especially, about when action becomes advocacy or, put differently, which forms of advocacy are palatable to SRIC and which are not. It remains to be seen whether this underlying tension or misalignment within the committee becomes overt and whether there will be spillover effects on the action-learning partnership with CUISR.

Navigating this ebb and flow of alignment requires patience and trust between the partners, but they might not be sufficient. In our case, we also required an informed, engaged intermediary. The development and work of the EWG have comprised the crucial turning point in ensuring that alignment outweighs misalignment. Consistent with the principles of CBR, it facilitates dialogic development of the research plan, manages co-production of new knowledge, spearheads knowledge translation, provides a forum for reflexivity, and in turn bridges the differing institutional realities of SRIC and CUISR.

We began this chapter with a question in the title. After the opportunity to trace our journey and to analyze it according to the basic tenets of CBR, our answer is, yes, the SRIC-CUISR collaboration is making strides toward conducting valid and important community-based research of the participatory action variety. To date, the action has focused primarily on SRIC and its member organizations, including CUISR, but it is having an impact, directly and indirectly, on how the members view and participate in participatory evaluation. The extent of the partnership's impact on policy change in the areas of poverty reduction and Aboriginal employment remains uncertain. One thing is certain, however; our future journey is bound to launch more questions, more exploration and innovation, and continued reflexivity on our progress.

Acknowledgements

We would like to acknowledge the Saskatoon Regional Intersectoral Committee for funding the evaluation project that spawned this interesting partnership. In particular, we want to thank the SRIC coordinator, Fred Ozirney, and fellow members of the Evaluation Working Group: Dorothy Hyde (Radius Community Centre), Crandall Hrynkiw (Ministry of Education, Central Region), and Doug Rain (Ministry of Advanced Education, Employment and Immigration and Co-Chair, EWG). Thanks also to SRIC Co-Chairs Bev Hanson (Greater Saskatoon Catholic Schools) and Garry Prediger (Ministry of Social Services) and past Co-Chair Chris Broten (Ministry of Advanced Education, Employment and Immigration) for their unwavering support for advancing SRIC.

Appendix A: SRIC Evaluation Framework

In early 2009, SRIC approved an evaluation framework presented by CUISR. It depicts SRIC as operating through three concentric rings to achieve its mission of improved human service delivery in the Saskatoon region: the SRIC members themselves, their affiliated ministries and agencies, and Saskatoon and area residents. The boundaries between the rings are permeable, with influences moving from the centre outward and from the citizenry inward.

When the poverty reduction and Aboriginal employment working groups were formed in 2009, they were added to the framework. Figure 1 below shows that the potential impact of the working groups spans all three rings, the relationship between SRIC and the Aboriginal employment group is a direct (i.e., controlling) one, and the relationship between SRIC and the poverty reduction group is an indirect one, with SRIC playing a supporting rather than directing role.

Research questions are associated with each of the rings, as follows.

Inner Ring: SRIC

- What is SRIC doing well?
- What is the current level and quality of coordination with respect to linkages, information sharing, databases, and funding?
- How has coordination changed over time?
- How can coordination improve?

Middle Ring: Affiliated Ministries and Agencies

- Where and how has SRIC had influence on policies, programs, and resource deployment; new coordinated program delivery; and/or reduction of service gaps or redundant overlap?

With the addition of the working groups, other questions were added.

- In what ways have the working groups affected SRIC's influence on policies, programs, and resource deployment?
- What needs for innovation and coordination have been identified by the working groups?

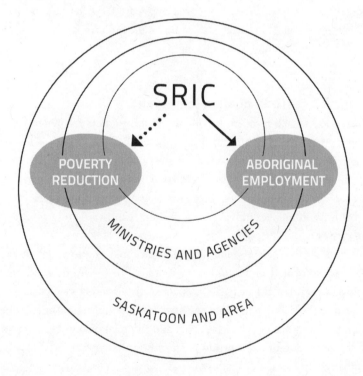

Figure 1. SRIC evaluation framework

- What processes have been put in place within SRIC and with its member ministries and agencies to facilitate that innovation and coordination?

Outer Ring: The Saskatoon Region
- How have outcomes for residents changed?
- How effective has SRIC been in addressing community priorities?
- What are future directions for SRIC?

A report on collaboration within SRIC has been published (Evaluation Working Group, 2009), and evaluation of the middle ring is being completed. It too will involve a discussion by SRIC members guided by the EWG, and the results will contribute to SRIC's next strategic plan.

References
Alkin, M. C. (1991). Evaluation theory development: II. In M. W. McLaughlin & D. C. Phillips (Eds.), *Evaluation and education: At quarter century* (pp. 91–112). Chicago: University of Chicago Press.

Bryk, A. S. (Ed.). (1983). *Stakeholder-based evaluation.* New Directions for Program Evaluation 17. San Francisco: Jossey-Bass.

Cousins, J. B., & Whitmore, E. (1998). Framing participatory evaluation. *New Directions for Evaluation, 80,* 87–105.

Elliott, P. (2011). *Participatory action research: Challenges, complications, and opportunities.* Saskatoon: Community-University Institute for Social Research and Centre for the Study of Co-operatives. University of Saskatchewan.

Evaluation Working Group. (2009). *Evaluation report: Snapshot of collaborative processes.* Saskatoon: SRIC.

Flicker, S., Savan, B., Kolenda, B., & Mildenberger, M. (2008). A snapshot of community-based research in Canada: Who? what? why? how? *Health Education Research, 23*(1), 106–114.

Garaway, G. B. (1995). Participatory evaluation. *Studies in Educational Evaluation, 21*(1), 85–102.

Heller, K. (1989). The return to community. *American Journal of Community Psychology, 17*(1), 1–15.

Lemstra, M., & Neudorf, C. (2008). *Health disparity in Saskatoon: Analysis to intervention.* Saskatoon: Saskatoon Health Region.

McMillan, D. W., & Chavis, D. M. (1986). Sense of community: A definition and theory. *Journal of Community Psychology, 14*(1), 6–23.

Roche, B. (2008). *New directions in community-based research.* Toronto: Wellesley Institute.

COMMUNITY-BASED RESEARCH THROUGH COMMUNITY SERVICE LEARNING
BENEFITS AND CHALLENGES FOR THE COMMUNITY AND UNIVERSITY

Nicola Chopin, Maria Basualdo, and David McDine

Introduction

The Community-University Institute for Social Research (CUISR) engages students, faculty, and community members in partnerships that result in relevant and responsive research to achieve social change. Our experience with community-based participatory action research (CBPAR) over the past decade has led us to rethink traditional research models and strategies by increasing flexibility, adopting extended timelines, investing additional resources, and gaining diverse input to contribute to community-building efforts. CUISR has also developed the guiding principles of research, relationships, and reflexivity (discussed in greater detail by Findlay, Ray, and Basualdo in this volume), which guide us from the first meetings of community-university partners to discuss research design and direction to final approval of strategies for dissemination.

CUISR's CBPAR approach employs research teams comprising a student, faculty member, and community partner, an approach established during CUISR's tenure as a Social Sciences and Humanities Research Council of Canada (SSHRC) Community-University Research Alliance (CURA), which emphasizes not only research but also academic benefits for students (SSHRC, 2010). This team approach can be conceptualized as community service learning (CSL), defined here as experiential learning that integrates community service and academic course-based learning (Furco, 1996; Howard, 1998; Mooney & Edwards, 2001; Siefer, 1998). CSL provides students with the opportunity to acquire new knowledge and skills, concretely apply abstract academic concepts, and provide a service that meets community needs (Bringle & Hatcher, 2000; Markus, Howard, & King, 1993; Parker-Gwin & Mabry, 1998; Reardon, 1998). Reflection, which encourages students to connect experiential and academic learning, is a critical component of CSL (Bringle & Hatcher, 2000; Hondagneau-Sotelo & Raskoff, 1994; Parker-Gwin & Mabry, 1998; Siefer, 1998).

Because of its emphasis on integrating experiential learning, community service, and course-based learning (Reardon, 1998), the CBPAR conducted at CUISR is well suited as an avenue for providing CSL (though not all CSL takes the form of CBPAR, for the purposes of this chapter CSL refers to CUISR's CBPAR approach). CUISR interns at the undergraduate, master, and doctoral levels most frequently conduct research as part of a practicum, cooperative, or thesis project, which necessitates applying course-based learning to their CUISR research projects. Because CUISR adopts a team approach, mentoring support is available from institute staff, faculty members, and community partners to build on academic learning through experiential learning. And since community partners request, conceptualize, and carry out research, the result is a project responsive to community needs. Finally, CUISR's reflexivity principle aligns with the CSL requirement of reflection.

This chapter describes the impact of CUISR's CSL program on its students, faculty members, and community partners. We present the results of two evaluations and discuss implications for the institute, community service learning, and community-based research. The chapter ends with a discussion of lessons that we have learned from ten years of providing CSL through CBPAR.

Background: Benefits and Challenges Associated with CSL

Research into CSL has focused primarily on student-related outcomes. Although there is a shortage of outcome research focusing on community and faculty, new conceptualizations of CSL have begun to emerge that emphasize more complex outcomes for students, faculty, and community (Gemmel & Clayton, 2009). These new conceptualizations recognize that

CSL provides the opportunity to develop multiple types of knowledge (e.g., academic, community based, Indigenous), contributions, skills, and perspectives. These types of knowledge create innovative opportunities for sharing, collaborating, and generating knowledge (Giles, 2008; Nyden, 2003).

Benefits of Community Service Learning

Evaluations of CSL programs have traditionally focused on outcomes for students (Gemmel & Clayton, 2009) although the literature on academic benefits is mixed. While some evaluations have found that CSL students earn higher grades (Astin, Sax, & Avalos, 1999; Markus et al., 1993), understand theoretical and course content better (Hynie et al., 2011; Siefer, 1998), and acquire practical skills (Hynie et al., 2011) than students in traditional curricula, other research finds little improvement in these outcomes (Parker-Gwin & Mabry, 1998) or only under specific circumstances. For example, Gray and colleagues (1999) found improvements in students' academic abilities only with "better" practices such as linking service with course content, providing training and supervisions, having students volunteer for at least twenty hours a semester, and having students reflect on service during class discussions. Key improvements have also been seen in students' professional skills such as leadership ability (Astin et al., 1999), interpersonal skills (Gray et al., 1999), problem-solving skills (Mooney & Edwards, 2001), critical-thinking skills (Astin et al., 1999; Mooney & Edwards, 2001; Parker-Gwin & Mabry, 1998), and conflict resolution and collaborative skills (Astin et al., 1999; Reardon, 1998).

Many scholars consider community-engaged scholarship to be a vital component of CSL (Bamber & Hankin, 2011; Norris-Tirrell et al., 2010; Swords & Kiely, 2010) and advocate CSL as a transformative experience that effectively increases students' civic engagement (Bamber & Hankin, 2011; Dirkx, 1998; Hynie et al., 2011; Millican & Bourner, 2011). As a result CSL can be a very effective community development tool (Swords & Kiely, 2010). Evaluations have found that students' participation in CSL is associated with increased awareness of social issues and community needs (Astin, Sax, & Avalos, 1999; Mooney & Edwards, 2001; Siefer, 1998), increased appreciation of social justice approaches (Bamber & Hankin, 2011), commitment to community and volunteerism (Astin et al., 1999; Markus et al., 1993; Siefer, 1998), and civic responsibility (Gray, Ondaatje, & Zakaras, 1999; Myers-Lipton, 1998). Ultimately, by exposing students to politically, socially, and economically complex issues for which there are no easy solutions, CSL can be a powerful tool for challenging students' existing worldviews (Bamber & Hankin, 2011; Hynie et al., 2011; Norris-Tirrell et al., 2010: Swords & Kiely, 2010). The result can be the creation and understanding of new knowledge, which serves to enable students to develop new meanings and worldviews (Bamber & Hankin, 2011; Dirkx, 1998; Hynie et al., 2011; Norris-Tirrell et al., 2010; Swords & Kiely, 2010).

There is a small body of literature examining CSL's impact on faculty and community partners. It provides faculty with opportunities to conduct community-responsive research (Reardon, 1998) and funding and publishing opportunities (Nyden, 2003). In addition, CSL has been found to be a "transformational" experience that can increase volunteerism among faculty (Siefer, 1998, p. 276). At an institutional level, it benefits post-secondary institutions by augmenting universities' impact beyond the confines of campus, particularly as services and community engagement are reflected in university strategic plans and missions (Hall, 2009; Hynie et al., 2011; Millican & Bourner, 2011). Community benefits include establishing reciprocal partnerships that support academic and community goals (Bringle & Hatcher, 2000; Siefer, 1998), which include opportunities to train students and increased capacity to provide services (Siefer, 1998). When CSL takes the form of CBR, the capacity for problem solving by community-based organizations increases, and recommendations coming out of the research are more likely to be implemented and sustained (Reardon, 1998). Students also offer community-based organizations increased labour and resources, including access to university resources and collaborations (Blouin & Perry, 2009), and they increase the capacity of community-based organizations to conduct research of interest to the organization (Rosing & Hofman, 2010).

Challenges of Community Service Learning

The literature highlights several challenges associated with CSL. Several authors note the importance of taking the time to develop lasting, mutually beneficial relationships, particularly since community-based organizations might mistrust "ivory tower" researchers (Gray et al., 1999; Maddux, Bradley, Fuller, Darnell, & Wright, 2006; Shrader, Saunders, Marullo, Benatti, & Mass Weigert, 2008). Although time-consuming, engaging community partners in identifying and defining research questions and processes, analyzing data, and interpreting results fosters stronger relationships (Giles, 2008; Maddux et al., 2006; Nyden, 2003; Reardon, 1998).

Balancing community and academic needs can also be challenging. As an approach to teaching, CSL must satisfy course objectives while also accommodating community needs (Maddux et al., 2006; Stoecker et al., 2010). The academic semester can cause difficulties for community-based organizations, particularly with long-term projects (Gray et al., 1999; Maddux et al., 2006; Rosing & Hofman, 2010; Shrader et al., 2008). The investment of working with and training students may at times be greater than the benefits obtained from the students' project (Stoecker et al., 2010).

Finally, a lack of faculty and institutional support acts as a barrier since some faculty members believe that CSL is a less rigorous teaching method (Giles, 2008; Howard, 1998; Maddux et al., 2006). CSL and related participatory research also tend not to be recognized in hiring, promotion, tenure, and merit decisions (Bringle & Hatcher, 1996, 2000; Nyden, 2003; Shrader et al., 2008; Swords & Kiely, 2010), despite the rich research, funding, and publishing opportunities provided by conducting CBPAR through CSL (Nyden, 2003).

Evaluating Community Service Learning at CUISR

CUISR has had a CSL program for ten years, during which time two evaluations of our impact as an institute have provided information to improve our CBPAR processes. These evaluations—a 2005 review of the institute's impact as a SSHRC-CURA and a 2010 evaluation of CUISR's CSL program—have contributed to the development of our CSL program.

2005 SSHRC Review of CUISR

In 2005, a review of CUISR's impact as a SSHRC-CURA on its stakeholders and larger community examined the institute's "contribution to building community capacity through the key stakeholders interested in improving the quality of life for the citizens of Saskatoon" (Sanderson, 2005, p. 8). The review focused on three areas of impact (students, community-based researchers, and the wider community) and examined CUISR's adherence to its guiding principles with regard to certain dimensions of successful community-based initiatives (Moote et al., 2001).

The 2005 review explored the benefits and challenges experienced by CUISR's students and community partners through interviews with twenty-seven community partner representatives, researchers, staff, and CUISR leaders; document reviews; and informal information gathering during brown bag luncheons and public forums; it also provided information on making improvements to our CSL program.

2010 Internal Evaluation of CUISR's CSL Program
In 2010, CUISR evaluated the impact of our CSL program on students, community researchers, and faculty members as part of its strategic priority to analyze community-university partnerships. The first round of data (reported here) was collected via an online survey from fourteen student researchers, four faculty members, and nine community partners during the summer of 2010. The questionnaire examined outcomes for students, faculty, and community partners identified from the literature and included satisfaction with the research process; benefits, uses, and impacts of the research and working with CUISR; contributions to skill development and civic attitudes; and areas for improvement. Faculty members were also asked about the relationship between CBPAR, the academic environment, and career opportunities. Based on the first round of results, the questionnaire will be improved and implemented as an exit survey, providing us with the ability to evaluate our progress quickly and implement changes on an ongoing basis.

Outcomes and Lessons Learned
Both projects identified benefits and challenges for students, faculty, and community partners, and the results provided information on CUISR's CBPAR approach that led to several improvements that directly impacted our CSL program.

Enhanced Learning, Quality Research, and Community-University Partnerships: 2005 Evaluation Findings
The 2005 review found that *enhanced learning* was a key impact for students working with CUISR. Specifically, working with the institute provided students with the opportunity to do qualitative research, which improved their networking and research skills (including interviewing, managing projects, applying methodologies, communicating, and thinking critically)—all while learning about their community. Students particularly valued the experience of writing and publishing academic reports. Many community partners thought that students made valuable contributions to the research and found their knowledge, input, and perspectives worthwhile. Many students also decided to pursue research related to their CUISR research in their academic programs, found employment with their project's community partners, and had strengthened commitments to conduct CBR as a result of working with the institute.

In addition, CUISR's ability to provide inexpensive, impartial, *academic-quality research* was a key benefit of working with CUISR. Community partners reported being able to collect valuable information on their community and their particular organization's needs, roles, and services, which allowed them to influence programs, activities, and policies. CUISR research was identified as a community development tool that allowed community partners to support and justify their programming and funding requests. Importantly, participants reported that their stakeholders and funders perceived CUISR research as more credible and responded more favourably to recommendations relative to in-house research.

Community partners were highly supportive of the institute's CBPAR approach, with many reporting that the research exceeded their expectations and was based on a true community-university partnership, which resulted in networking benefits, including meeting and working with other community-based organizations and academics.

Academic Opportunities, Employability, Civic Benefits, and Relationships with Community: 2010 Evaluation Findings

Student, faculty, and community partner rating of their satisfaction with working with CUISR was high, with an average satisfaction rating of 4.3 (students), 4.5 (faculty), and 4.7 (community partners) out of 5.

The evaluation identified several *academic benefits* for students and faculty. Students valued the educational benefits of their CUISR internships, believing that they enhanced their university education and increased their awareness of biases or prejudices. Students thought that their internships allowed them to develop or refine their interpersonal, communication, and collaboration skills as well as their research skills (e.g., interviewing, writing, presenting, project and time management, and tailoring reports to non-academic audiences). They also valued the opportunity to apply theory to practice and gain a greater understanding of the complexities of social problems. Faculty respondents agreed that their CUISR research projects improved students' ability to apply academic course material to real-world problems while also providing faculty with opportunities to conduct interdisciplinary research not normally available within their colleges. Community partners felt a responsibility to provide training opportunities for students and thought that their contributions to student training were beneficial.

Students thought that their CUISR internships increased *employment readiness* by providing them with work experience that enhanced their resumés and gave them experience that set them apart from their peers. Community partners were highly satisfied with their interns' professionalism, including respectfulness, reliability, level of training, ability to meet their particular organization's needs, and quality of work. Half of the students surveyed indicated that their experiences with CUISR influenced their career aspirations.

In terms of *civic and community benefits*, students believed that their internships gave them meaningful ways to contribute to their communities and research areas. Students felt strongly that they had a responsibility to serve their communities and identified their internships as contributing to their understanding of their communities' needs and problems as well as their roles as citizens. Both faculty and students thought that their CUISR research was beneficial, contributing to improving quality of life in the community. Community partners cited many benefits of working with the institute, including increased capacity and access to high-quality, actionable research—uses of the CUISR research included education, presentations, and advocacy with politicians and ministers, as well as information to inform decisions and change work plans, procedures, programs, and strategic plans. Community partners also believed that working with CUISR allowed them to raise their profiles through research. One partner appreciated CUISR's ability to balance academics and pragmatics when conducting research.

Finally, community partners indicated the importance of *CUISR's reputation and relationships* in producing actionable and relevant (but low-cost) CBR. Specifically, the institute's commitment to community-university partnerships; familiarity and connections with local groups, the community, and the university; focus on community development; and strategic directions were identified as reasons why community partners became involved with CUISR—and reasons that their organizations would continue to work with CUISR in the future.

Lessons Learned

The 2005 evaluation highlighted six key domains for improvement. The evaluation identified the need for improvement related to *team development*, which included defining and documenting all research team members' responsibilities at the planning stages, with some community partners desiring greater participation in the research. The need for *improved communication* among CUISR, community partners, and students was also raised, with one community partner being disappointed in communication during its project. Recommendations included devoting time to developing strong working relationships between students and community partners to allow the student researcher to get to know the organization, establish communication procedures, educate the community partner on the research process, and provide time to disseminate research findings.

Research projects often took longer to complete than initially expected. The need to adopt *extended timelines* was associated with the complexities of conducting CBR, in which unexpected issues requiring resolution and renegotiation often arise. The evaluation also found difficulties related to finalizing reports within the project timeline, and this was related to difficulties with having academic supervisors and community partners review reports in a timely manner. Adopting extended timelines was identified as beneficial for evaluating, debriefing, reflecting on, and learning from the research, and it has improved team development and ensured that proper time is allocated to completing research.

A need for *clarity of roles and responsibilities* was identified. Several community partners and students were unclear whether CUISR or the community partner was responsible for supervising the student (particularly where disciplinary and interpersonal difficulties were concerned). Several community partners thought that the faculty member could have played a more active role in the research. Finally, some community partners discussed issues related to ownership and control of the research findings, with some interviewees being concerned about CUISR having sole authority over, control of, and credit for conducting the research. As a result, issues related to ownership are currently discussed during planning stages.

A primary goal of CUISR is to train students to conduct CBR, and the evaluation found a need for community partners or faculty members to orient, support, and *mentor students* to ensure that students have the skills necessary for collecting data and interacting with clients. A strong relationship with the faculty supervisor was identified as an important factor for enhancing learning and completing projects. Several student researchers were concerned about having regular or sufficient access to the project's academic supervisor and thought that having more support was necessary. As a result, a regular practice of providing student interns with an orientation package and holding regular meetings with students to provide support as needed was implemented.

The evaluation highlighted the need for more *effective knowledge transfer* mechanisms. Some community partners thought that knowledge transfer between the university and the community was primarily one way. Lack of time for debriefing and other forms of knowledge transfer contributed directly to the difficulties experienced with knowledge transfer. This finding is extremely important given CUISR's philosophy of community-university partnership. Efforts are currently devoted to ensuring that products coming out of the research are tailored to the intended audiences, and students, faculty, and community partners are encouraged to present research findings at community organizations, brown bag lunches, conferences, and community forums as appropriate.

In 2010, participants identified key areas for successful project completion and provided suggestions for improvement. Many lessons learned were associated with *project planning and development*. Students identified being involved with project development and communication as important for success of the project. Students were relatively less satisfied with the partnership formed between the community partner and the academic supervisor. They also identified regular check-ins with and support from CUISR staff as important, as was communication between the student, CUISR staff, and the community partner. Community partners believed that CUISR projects were most successful when roles, responsibilities, timelines, and deliverables were specified during planning. One community partner noted the importance of proper communication for developing the relationship among students, faculty member, and community partners. A faculty member recognized that CUISR research required extended timelines relative to other types of research.

A key finding of the 2010 evaluation was a need for *greater support* for research teams. Students, on average, were satisfied with the support provided during their internships and thought that earnest attempts were made to answer their questions or concerns; their satisfaction was lower (with average scores of 3–4 out of 5) with instruction preparing them to complete their internship, amount of work required, feedback on their performance, and clarity of work required. Students also felt a need for greater monetary compensation. They suggested that more support be provided in terms of templates and orientations regarding ethics applications, focus groups, and qualitative research, suggestions that are supported by the relatively lower ratings that these areas received in the survey. Faculty, on average, were satisfied with the amount of work required of them (relative to other areas), though students felt the need for a stronger relationship with their academic supervisors. Faculty were also less likely to endorse that their experiences with CUISR allowed them to display how their work is relevant to the larger community. Finally, frequency of communication received a relatively lower rating on the survey from community partners.

Faculty were neutral or dissatisfied in their ratings of the *academy's support for CBR*, assigning lower ratings to the value that their college placed on CUISR research as well as the support received from colleagues, administrators, and the tenure and promotion system to do research with CUISR. These faculty barriers indicate that institutionalization of CSL and CBR might be problematic at the University of Saskatchewan. A lack of institutionalization might also be reflected by one community partner's suggestion that the institute find sustainable funding to continue its work.

Next Steps: Overcoming Barriers

When examining the results of the two evaluations, it is apparent that CUISR's experience with CSL is consistent with findings from other CSL research. The evaluations highlighted the academic, employment, and civic benefits that students experience from their internships. Community benefits include access to actionable and relevant research that assists in the development of community and community-based organizations. CUISR research also provides faculty with the interdisciplinary research opportunities not readily available to them.

Challenges that CUISR faces in its CSL program echo barriers identified in the literature. For example, the need to adopt extended timelines is a general finding (e.g., Maddux et al., 2006) and was a key learning that became part of CUISR's guiding principles in 2005. This barrier, identified as a requirement for developing strong relationships, dealing with unforeseen circumstances, and finalizing reports in 2005, was not identified directly as an issue in 2010; however, the results of the 2010 evaluation suggest that more time could be spent fostering the relationship among community partners, students, and faculty members. As a result, communication—identified as an area for improvement in both evaluations and a key difficulty noted in the literature (Gray, Ondaatje, & Zakaras, 1999)—is a high priority.

The need to define and document roles and responsibilities at the beginning of a project, noted in the literature review (Shrader et al., 2008) and recommended by a community partner in the 2005 evaluation, was advanced in 2010 by a community partner as a key learning over the course of its partnership with CUISR. The results of the 2010 evaluation suggest that, although faculty are clear on their roles and responsibilities, students can benefit from additional clarification regarding expectations (however, this item's score was 3.3 out of 5, suggesting that satisfaction was still relatively high).

Moreover, the need for increased student training support was identified in 2005 and 2010, although we saw some change between the two evaluations. In 2005, this area was identified by a community partner concerned about students' ability to work with its clients; in 2010, students requested more support and orientation in creating ethics applications, delivering focus groups, and conducting qualitative research.

Concerns related to ownership of findings and one-way knowledge translation were found in 2005 and are extremely important in light of CUISR's philosophies. This key difficulty is part of a larger problem of restricting the community's voice in academic research that can lead to the community not being involved in identifying and defining research questions, analyzing data, and interpreting findings (Giles, 2008; Nyden, 2003; Reardon, 1998). Concerns relating to ownership were not raised in 2010, suggesting that some improvements have been made in this area, although faculty were less satisfied that their experience with CUISR allowed them to display how their work is relevant to the larger community.

Finally, faculty members indicated that their work with CUISR is not recognized in tenure and promotion decisions, although only four faculty members responded to the survey. Despite the small number of respondents, a pervasive lack of institutionalization of CSL is noted in the literature (Bringle & Hatcher, 1996, 2000; Nyden, 2003; Shrader et al., 2008), which suggests that the results might be more than anecdotal. Changes made at other institutions include modifying tenure and promotion practices and providing time and grants for faculty to participate in CBR and CSL (Nyden, 2003; Shrader et al., 2008). Despite current hurdles for faculty engaged in CBR, this area is increasingly recognized by institutions as being important. CUISR, its board members, and affiliated faculty can play a role—locally, regionally, and nationally—in advocating for increased recognition of community-engaged scholarship.

Team building, student training, knowledge translation, and faculty engagement are currently priorities for improvement at CUISR. Some of this support—particularly providing orientations, background materials on CBR, and templates and samples of work plans, ethics applications, and other project materials—can be provided by institute staff. However, the greatest potential for solving many of the issues identified above appears to lie in improving faculty and community partner support for students. As noted above, a lack of recognition within the academy regarding the merits of CBR (and CSL as an approach to teaching) has played a role in limiting the number of faculty members engaged with CUISR. Currently, our strategy is to expand our network of faculty members to ensure that they have programs of research that are strong fits with community partners' research needs and students' interests. In addition, CUISR is adopting an increased focus on fostering relationships within research teams; we are currently experimenting with a "science shop" approach to some of our projects. In this approach, institute staff act as a hub to foster contacts among students, faculty, and community partners; preliminary results suggest that this approach might yield stronger relationships among team members.

Finally, the 2010 evaluation highlighted an area that is related less to the functioning of research teams than to the larger context in which CUISR research occurs: one community partner suggested that CUISR develop sustainable funding. The institute currently does not receive core funding from the University of Saskatchewan, and it is focused on addressing this area by applying for large academic and government grants on an ongoing basis. Prior research on CSL highlighted the importance of having a centralized office to provide administrative support, mentoring, and training opportunities for those involved with CSL (Bringle & Hatcher, 2000; Giles, 2008), with some literature suggesting that, in practice, more funding often goes to developing a sustainable infrastructure in the community and the institution than directly providing CSL. Sustainable and core funding could contribute greatly to the operations of CUISR, bolstering its ability to continue its work into the future.

Conclusion

The evaluations of CUISR's CSL program highlight the many benefits of CUISR's approach for students, faculty members, and community partners. These evaluations provide evidence that delivering CSL through CBR is an effective teaching method that contributes not only to student education and skill development but also to faculty's access to interdisciplinary research and community partners' ability to produce rigorous academic research that responds to their needs.

Our case echoes many of the strengths and difficulties associated with conducting CSL and CBR, providing many important lessons learned over the past ten years. Adopting extended timelines and establishing strong communications to develop meaningful, lasting partnerships are important lessons and have contributed greatly to CUISR's success. We have also learned that a high level of training for and mentoring of all research partners is required to ensure positive outcomes. As the institute responds to the 2010 evaluation findings, we will continue to develop and refine our approach to ensure that our academic and local communities continue to benefit from socially responsive, rigorous, and actionable research.

Acknowledgements

We would like to thank SSHRC for funding much of the research evaluated here. We would also like to thank CUISR for supporting the development and writing of this chapter. Thank you to all of the community partners, faculty members, and students who contributed to the research discussed here.

References

Astin, A. W., Sax, L. J., & Avalos, J. (1999). Long-term effects of volunteerism during the undergraduate years. *Review of Higher Education, 22*(2), 187–202.

Bamber, P., & Hankin, L. (2011). Transformative learning through service-learning: No passport required. *Education & Training, 53*(2/3), 190–206.

Blouin, D. D., & Perry, E. M. (2009). Whom does service learning really serve? Community-based organizations' perspectives on service learning. *Teaching Sociology, 37*(2), 120–135.

Bringle, R. G., & Hatcher, J. A. (1996). Implementing service learning in higher education. *Journal of Higher Education, 67*(2), 221–239.

Bringle, R. G., & Hatcher, J. A. (2000). Institutionalization of service learning in higher education. *Journal of Higher Education, 71*(3), 273–290.

Dirkx, J. M. (1998). Transformative learning theory in the practice of adult education: An overview. *PAACE Journal of Lifelong Learning, 7*, 1–14.

Furco, A. (1996). Service-learning: A balanced approach to experiential education. In *Expanding boundaries: Service and learning* (pp. 2–6). Washington, DC: Corporation for National Service.

Gemmel, L. J., & Clayton, P. H. (2009). *A comprehensive framework for community service learning in Canada*. Ottawa, ON: Canadian Alliance for Community Service Learning.

Giles, D. E. (2008). Understanding an emerging field of scholarship: Toward a research agenda for engaged, public scholarship. *Journal of Higher Education Outreach and Engagement, 12*(2), 97–106.

Gray, M. J., Ondaatje, E. H., & Zakaras, L. (1999). *Combining service and learning in higher education: Summary report.* Santa Monica, CA: RAND.

Hall, B. L. (2009). Higher education, community engagement, and the public good: Building the future of continuing education in Canada. *Canadian Journal of University Continuing Education, 35*(1), 11–23.

Hondagneau-Sotelo, P., & Raskoff, S. (1994). Community service-learning: Promises and problems. *Teaching Sociology, 22*(3), 248–254.

Howard, J. P. F. (1998). Academic service learning: A counternormative pedagogy. *New Directions for Teaching and Learning, 73*, 21–29.

Hynie, M., Jensen, K., Johnny, M., Wedlock, J., & Phipps, D. (2011). Student internships bridge research to real world problems. *Education & Training, 53*(2/3), 237–248.

Maddux, H. C., Bradley, B., Fuller, D. S., Darnell, C. Z., & Wright, B. D. (2006). Active learning, action research: A case study in community engagement, service-learning, and technology integration. *Journal of Higher Education Outreach and Engagement, 11*(3), 65–79.

Markus, G. B., Howard, J. P. F., & King, D. C. (1993). Integrating community service and classroom instruction enhances learning: Results from an experiment. *Educational Evaluation and Policy Analysis, 15*(4), 410–419.

Millican, J., & Bourner, T., (2011). Editorial: Student-community engagement and the changing role and context of higher education. *Education & Training, 53*(2/3), 89–99.

Mooney, L. A., & Edwards, B. (2001). Experiential learning in sociology: Service learning and other community-based learning initiatives. *Teaching Sociology, 29*(2), 181–194.

Moote, M. A., Brown, B. A., Kingsley, E., Lee, S. X., Marshall, S., Voth, D. E., & Walker, G. B. (2001). Process: Redefining relationships. *Journal of Sustainable Forestry, 12*(3–4), 97–116.

Myers-Lipton, S. J. (1998). Effect of a comprehensive service-learning program on college students' civic responsibility. *Teaching Sociology, 26*(4), 243–258.

Norris-Tirrell, D., Lambert-Pennington, K., & Hyland, S. (2010). Embedding service learning in engaged scholarship at research institutions to revitalize metropolitan neighborhoods. *Journal of Community Practice, 18*, 171–189.

Nyden, P. (2003). Academic incentives for faculty participation in community-based participatory research. *Journal of General Internal Medicine, 18*, 576–585.

Parker-Gwin, R., & Mabry, J. B. (1998). Service learning as pedagogy and civic education: Comparing outcomes for three models. *Teaching Sociology, 26*(4), 276–291.

Reardon, K. M. (1998). Participatory action research as service learning. *New Directions for Teaching and Learning, 73*, 57–64.

Rosing, H., & Hofman, N. G. (2010). Notes from the field: Service learning and the development of multidisciplinary community-based research initiatives. *Journal of Community Practice, 18*, 213–232.

Sanderson, K. (2005). *Community-University Institute for Social Research: Partnering to build capacity and connections in the community.* Saskatoon: CUISR.

Shrader, E., Saunders, M. A., Marullo, S., Benatti, S., & Mass Weigert, K. (2008). Institutionalizing community-based learning and research: The case for external networks. *Michigan Journal of Community Service Learning*, (Spring), 27–40.

Siefer, S. D. (1998). Service-learning: Community-campus partnerships for health professions education. *Academic Medicine, 73*(3), 273–277.

Social Sciences and Humanities Research Council of Canada (SSHRC). (2010). *Framing our directions 2010–12: Social Sciences and Humanities Research Council of Canada.* http://www.sshrc-crsh.gc.ca/.

Stoecker, R., Loving, K., Reddy, M., & Bollig, N. (2010). Can community-based research guide service learning? *Journal of Community Practice, 18*, 280–296.

Swords, A.C.S., & Kiely, R. (2010). Beyond pedagogy: Service learning as movement building in higher education. *Journal of Community Practice, 18*(2/3), 148–170.

TRIPARTITE COLLABORATION AND CHALLENGES
REFLECTING ON THE RESEARCH PROCESS OF A PARTICIPATORY PROGRAM EVALUATION

Hongxia Shan, Nazeem Muhajarine, and Kristjana Loptson

Introduction

The existent literature on community-based participatory research (CBPR) and participatory evaluation focuses extensively on establishing collaborative relationships between university-based researchers and local community groups (Brown & Vega, 1996; Jones & Wells, 2007; Spingett & Wallerstein, 2008; Williams, Labonte, Randall, & Muhajarine, 2005). In particular, emphasis has been placed on how researchers should position themselves in terms of approaching and working with the community for a meaningful and collaborative relationship. Special attention has been paid to the power differences between academic researchers and the community given the hierarchical ways in which their respective knowledge is socially recognized (Green & Mercer, 2001; Letiecq & Bailey, 2004; Sullivan et al., 2001; Wallerstein, 1999). The literature further shows that, without enough sensitivity to the power issue within the community, CBPR researchers can inadvertently reinforce the existing power relations to the detriment of vulnerable members of the community (Botes & van Rensburg, 2000; Njoh, 2002).

Our participatory evaluation of *KidsFirst*, a flagship early childhood intervention program in Saskatchewan, involves not only researchers and community-based organizations but also a government unit that has funded and directly overseen the program and commissioned the evaluation. Given the different social and power positions occupied by research partners, the tripartite partnership poses some challenges as well as presents some unique opportunities. This chapter adds to the CBPR scholarship with a reflective note on how to engage collaborators with converging yet diverging perspectives in an evaluative research project. Specifically, we revisit the process of the participatory evaluation, share the collaborative working strategies and outcomes of the researchers, and discuss the ways in which they reconciled divergent views and evidence.

The chapter is divided into five sections. Following the introduction, we briefly review the literature on CBPR and that on participatory evaluation to situate the chapter. We then provide some background information on the *KidsFirst* program and introduce how the participatory evaluation research came about in the first place. The fourth section focuses on the collaborative research process, particularly the collaborative strategies of the researchers, the challenges that they faced, and the lessons that they learned. We conclude with a recap of the chapter and a discussion of the implications of the research for other community-based researchers.

CBPR and Participatory Evaluation

In the past few decades, community-based participatory research has increasingly been adopted as a viable means to address health disparities and inequities (Israel, Eng, Schulz, & Parker, 2005; Minkler & Wallerstein, 2008; Wallerstein & Duran, 2006). CBPR is "a partnership approach to research that equitably involves, for example, community members, organizational representatives, and researchers in all aspects of the research process and in which all partners contribute expertise and share decision making and ownership" (Israel et al., 2005, p. 5). Rooted in action research (Reason & Bradbury, 2006), participatory methodologies (Green et al., 1995; Israel, Schulz, Parker, & Becker, 1998; Minkler & Wallerstein, 2008), and community development (Hall, 1984), and influenced by the work of Freire and some feminists of colour such as bell hooks (Wallerstein & Duran, 2006), CBPR constitutes a significant departure from the traditional paradigm of applied research in which the outside researcher largely determines the questions asked, the research instruments employed, and the kind of results and outcomes documented and valued (Gaventa,

1993). Instead, it is a research orientation (Cornwall & Jewkes, 1995) that emphasizes building trust, sharing power, fostering co-learning, enhancing strengths and resources, building capacity of research partners, and examining and addressing community-identified needs and health problems (Israel et al., 2005, p. 10). At the centre of the participatory orientation to evaluation and research is a shift of researchers' attitudes, "which in turn determines how, by, and for whom research is conceptualized, conducted," and "the corresponding location of power at every stage of the research process" (Cornwall & Jewkes, 1995, p. 1667).

"Program evaluation is the systematic collection of information about the activities, characteristics and outcomes of programs to make judgments of the program, improve program effectiveness and/or inform decisions about future programming" (Patton, 2002, p. 10). Participatory evaluation is an attempt to "involve in an evaluation all who ha[ve] a stake in its outcomes, with a view of these individuals and organizations' taking action and effecting change on the basis of the evaluation" (Spingett & Wallerstein, 2008, p. 200). Participatory evaluation can become part of the CBPR paradigm especially when research issues and questions originate with the community (Puma, Bennett, Cutforth, Tombari, & Stein, 2009). Proactive participatory evaluation shares with CBPR a common interest in research partnerships. In such a partnership, research stakeholders should be involved as research partners at every stage of the research process and retain ownership of the evaluation process and outcomes (Patton, 1997).

Clearly, most researchers engaged in CBPR and participatory evaluation are interested in collaborative research partnerships. Yet the type of collaboration that researchers focus on is often between university-based researchers and the non-academic community (Israel et al., 2005; Minkler & Wallerstein, 2008; Smith & Bryan, 2005; Wallerstein, 1999). Only a few studies have started to pay more attention to the intricacies of building partnerships with different stakeholders, including community organizations and policy-makers (Muhajarine et al., 2011; Williams et al., 2005). This chapter contributes to the existent literature by focusing on the challenges and accomplishments that some university-based researchers had when working with both community-based organizations and a government unit that funds and oversees the program, two groups of stakeholders with strong and specific interests in a participatory evaluation research project.

KidsFirst and the Tripartite Evaluation Research Partnership

KidsFirst is a flagship early childhood intervention program implemented in nine sites in Saskatchewan with high community needs. The program provides services and supports to vulnerable families with children up to five years old. It was launched in 2002 following the First Ministers' Early Childhood Development Agreement, which allocated funding to each province for the purpose of improving early childhood supports (Saskatchewan Education, Health, Intergovernmental and Aboriginal Affairs and Social Services, 2002). *KidsFirst* aims to promote the healthy growth and development of vulnerable children by bridging gaps in service delivery, removing barriers that keep families from accessing services, and mentoring families on parenting and life skills. The main mechanism of program delivery is through home visitation, in which trained paraprofessionals from partnering agencies meet regularly with families to assist them in meeting their basic needs, provide information, and encourage improvements in parent-child interactions. At the time of program intake, families are assessed and classified as low, medium, or high (complex) needs based on numerous variables, such as level of education, mental health, financial stability, substance addiction, and various other risk factors.

In 2006, the principal investigator of the *KidsFirst* evaluation project, Nazeem Muhajarine, was approached by the Ministry of Education of the Saskatchewan government to evaluate the short- and intermediate-term effectiveness of the program. (Prior to this contact, he had consulted with the ministry regarding development of data tools and measurement for the program in its early stages in 2002 and 2003.) Concurrently, an opportunity arose for the principal investigator to apply for external peer-reviewed funding from the Canadian Population Health Initiative (CPHI) for intervention research. While the proposal to CPHI was under consideration, the Ministry of Education funded development of the evaluation framework. CPHI granted funding for the quantitative portion of the evaluation, while the ministry and the nine *KidsFirst* sites provided funds for the qualitative portion of the evaluation (Muhajarine et al., 2011).

To maximize the reliability and utility of the program, a participatory approach was taken toward the evaluation, which involved both the provincial government staff and the local program staff. For this evaluation, a research team was formed that involved university-based researchers, government staff from the Early Childhood Development Unit (ECDU) at the Ministry of Education, which is responsible for the administration of

KidsFirst, and the director of evaluation in the Ministry of Social Services. The government assigned a half-time position as part of an in-kind contribution to the project. It also established an interministerial evaluation advisory committee made up of senior officials from the four ministries involved in *KidsFirst* (Health, Social Services, Education, and First Nations and Métis Relations). Program managers, though not part of the research team, were engaged as collaborators of the study; the Program Managers' Committee, an existing committee comprised of all nine *KidsFirst* program managers, participated in the evaluation process.

Reflecting on the Collaborative Process

In this participatory evaluation research, university researchers engaged both government and community-based partners throughout the study. In this section, we share the working principle and collaborative strategies of the researchers as well as their major challenges and the lessons that they learned while working with both community organizations and government partners, whose perspectives on and interests in the evaluation research differed at times.

Collaborative Principle and Strategies

The foundational principle of the participatory evaluation of *KidsFirst* was to engage the stakeholders in the research process as equal partners as well as co-learners. The researchers recognized that both partners brought in expertise, knowledge, and resources that should be valued. Although researchers are necessarily the instrument of research (Denzin & Lincoln, 2000, p. 368) in the sense that they are trained in the conduct of research (to collect, manage, and interpret data), an evaluation cannot be fully accomplished without contributions from other partners. It is important for researchers to "check their egos at the door" and adopt a learning posture throughout the research process. At the same time, it is important to facilitate learning about research among non-academic partners so that they can be informed and conscious participants in the study and shape the research process to maximize their ownership of the study.

To engage the community organizations and the policy-makers in the research process as equal partners and co-learners, we found three strategies most effective. The first is regular meetings and communications; the second is mobilizing the practical and experiential knowledge of partners;

and the third is co-authorship and formal acknowledgement of the contributions made by research partners in all published documents.

Constant communications and meetings are essential for a research project involving multiple partners. In our study, the research team met in person or by teleconference several times a year to guide the research, with meetings of smaller groups of team members occurring more often as needed. The interministerial evaluation advisory committee also met with the team members to advise and guide the evaluation, especially during the initial period of the study. Members of the research team met four to six times a year with the Program Managers' Committee and ECDU staff. In between these meetings, the research team also communicated with the program managers and ECDU to offer information, seek feedback, and give advice. Communication keeps information flowing three ways and is a crucial mechanism for all partners to provide and receive input and to learn about one another while working toward the common goal of understanding the effectiveness of the *KidsFirst* program.

The second strategy conducive to a collaborative relationship is to engage partners equally in a co-learning process (Minkler, 2000). Researchers need to mobilize and learn from the policy and practical knowledge of the partners for the substantive development of the project. For instance, when the team developed the evaluation framework and the program logic model, both policy-makers and program managers were involved in significant ways. The researchers started by laying out a prototype based on their review of the program manual and of materials from similar programs. They then held a workshop and small-group interactions with both policy-makers and program managers to map the core activities and desired program outcomes. Subsequently, the whole group had extensive back-and-forth e-mail communications in which the research partners provided feedback to the logic model until all three partners reached consensus.

By engaging both policy-makers and community program managers as equal partners, we do not mean that they were necessarily involved in the research process in the same manner and equally contributed to each component of the study. Instead, both partners were valued for their different knowledge of the program that comes with their specific roles in and experiences with it. For some components of the study, such as developing the evaluation framework, the policy-makers played a more central role in shaping the work, while the program managers provided useful feedback. In contrast, when the researchers developed the community profiles, the program managers, who had first-hand knowledge of their respective communities, were central to this work. They were co-authors of the community profiles report along with the relevant research team members. The principle at play here is that, throughout the research process as a whole, the researchers tried to maximize both partners' voices in the research and in ownership of the evaluation by mobilizing their different knowledge.

Throughout the study, the researchers also actively shared their expertise in and knowledge of the technical aspects of doing research. For instance, they shared their knowledge of research methods as well as their responsibility to keep confidential the personal information of research participants, which increased the willingness of the program managers to participate in the study and help to reach out to their clients. Additionally, the researchers directly contributed to the program by supplementing it with a theoretical foundation. *KidsFirst* follows a set of principles and philosophies. However, it lacked a theoretical support. The researchers, through a review of theories, helped to identify three theories that were consonant with the program: attachment theory[1], self-efficacy theory[2], and human ecology theory[3]. This theoretical review was communicated to the decision makers and program staff, who provided feedback and suggested ways to make the writing brief and concise so that it would be accessible to a larger audience. This particular theory review also helped to establish a common language for researchers, program funders, and organizational staff and facilitated their knowledge exchange in the evaluation process.

The third strategy, also the culmination of a productive collaborative relationship, is the practice of co-authorship, crediting partner contributions, and taking public ownership of the products. Throughout the research process, we encouraged all our research partners to be involved

• • • • • • • • •

1 Attachment theory proposes that mother-infant attachment patterns are foundational to the personality development of infants (Bretherton, 1992).
2 Human ecology theory focuses on humans as "both biological organisms and social beings in interaction with their environment." In this theory, "the family is considered to be an energy transformation system that is interdependent with its natural physical-biological, human-built and social-cultural milieu" (Bubolz & Sontag, 1993: 419).
3 Self-efficacy theory notes that an individual's self-efficacy level, or perception of "how well one can execute courses of action required to deal with prospective situations" (Bandura, 1982, p. 122) strongly influences the power an individual has in facing up with challenges and the actions one takes in challenging situations.

in producing research documents and to credit their contributions accordingly. For all documents produced from the project, the lead writers were made the main authors, followed by others who made significant contributions. In each document, members of the research team, research staff, and decision makers in the Ministry of Education and *KidsFirst* sites who were not named as authors were credited in the acknowledgements. For instance, the principal investigator, together with the research team, has co-written with the policy-makers an article on the process by which a partnership was formed for the study (Muhajarine et al., 2011). As well, researchers and program managers collaborated fully in the development of community profiles. Program managers actively participated in creating profiles of their respective sites. They provided materials and pictures and participated in the write-up. They were fully credited as the authors of this important document. These community profiles provided significant contextual knowledge of the program.

Challenges and Lessons Learned

Although the research proceeded smoothly overall, with full support from both partners, we encountered some challenges when finalizing the research findings. These challenges might have arisen largely because of the different standpoints and interests of the research partners, which reflected differences in authority and perhaps even resistance. Such differences in interest and authority can come into sharp focus in qualitative studies, in which voices, perspectives, and meanings matter much more than numerical data and their interpretation. They can also arise given the inclusive nature of a participatory evaluation, in which participation from very different perspectives, roles, and interests can shed different light on the program. While dealing with these differences, we learned that researchers need to give full consideration to all data received and fine-tune research findings to reflect different perspectives. We also learned that open dialogues and formal opportunities for conversations with all parties are crucial in producing inclusive findings and can lead to new questions that deserve attention from all stakeholders of the evaluation.

Traditionally, program evaluation is conducted to hold people accountable for policy goals and objectives (Vedung, 2004). As already mentioned, the intention of the government partner who commissioned the evaluation research was to understand whether and how the program has met its goals. The evaluation research was intended for management and administrative purposes if not as a monitoring tool. However, once the researchers started collecting people's experiences of the program, participants, especially staff from the community organizations, also tried to hold the government partner accountable for its "capacity to support [them] in realizing their potential and aspiration"; in other words, while relating their work and lived experiences, research participants tended to "return scrutiny to the sponsoring administrative system" (Kushner, 2005, p. 120) and to direct attention to how the system has enabled or inhibited their effectiveness at work.

Another issue that complicates the research process, but also makes participatory evaluation interesting and powerful, is that even the same individuals might relate different ideas, or points of view, at various stages of the research. Some CBPR researchers (Chávez, Duran, Baker, Avila, & Wallerstein, 2008) have pointed out that research participants can present researchers with a "public transcript" that can speak directly to the evaluation project. In addition, there can be a "hidden transcript" of what community members really think that remains outside the purview of researchers. As relationships and trust between researchers and communities evolve, hidden transcripts can be turned into public discourses. The newly emerged information, when tapped into, can increase the usefulness and relevance of the research findings.

Due to the different standpoints of the research partners and the developmental nature of participatory evaluation, the researchers were presented with contradictory and additional data, especially toward the end when the research findings were being finalized. While dealing with these challenges, we learned three lessons. The first lesson is that researchers need to treat contradictory data with care. In our study, the community and government partners clearly indicated different interests and views on certain issues. When researchers treat different data with serious consideration and care, further questions can arise that warrant attention from all parties involved. For instance, research participants, particularly program staff, reported a lack of training. Yet, when we shared this preliminary finding with the research partners, provincial staff proffered contradictory evidence; documented in their managerial data were a great number of training hours. When we brought the voices and numbers together, we became aware that the question went far beyond whether there was a lack of training. We were prompted to ask whether the content and method of training were relevant to the daily work carried out by home visitors and whether the training programs were offered at levels that suit home visitors with low literacy skills and without facility using computers. We were also led to question whether training was the solution to some of the issues raised. For instance, one of the problems reported by the home visitors was that they were not adequately trained or prepared for the high-risk crisis scenarios that they regularly faced in their work. We started questioning if paraprofessionals are best equipped to deal with complex-needs families.

The second lesson is that researchers must use care in representing stakeholders' voices and perspectives. This is especially important for studies involving multiple stakeholders. In the preliminary research findings, communicated to both the government and the community partners, we presented that the home visitors were underpaid and their turnover rate at some sites was high. These findings emerged from our interviews with not only home visitors but also program managers and Program Managers' Committee members. The government partners, when presented with these findings, criticized the subjective nature of the terms used, such as "underpaid" and "high." To counter these findings, they also offered administrative information detailing the salary scale of the home visitors as well as the turnover record of home visitors. Clearly, the same results were interpreted and understood in different ways by key partners in the study. After much deliberation, we retained these findings from the community, as significantly altering or losing them would threaten the integrity of the research process. We also understood, however, that beyond simple reporting of the interview findings we had an opportunity to play a greater role by altering word choices. We fine-tuned the report and made it clear that these were opinions expressed by staff members at some local sites.

The third lesson is that an open space of communication is necessary to produce inclusive, useful, and reliable research results. This was important even when we were made independent judges of the program. For instance, when it came to producing research recommendations, the principal investigator invited both partners to be involved. For some reason, they declined the invitation and decided that the researchers should make their independent recommendations. This independent space allowed by the research partners made it all the more important for us to make full use of meeting occasions and to digest their comments and feedback. The manner in which we came up with the recommendation regarding complex-needs family was a good example.

In the study, interviews with both program staff and program managers consistently showed that complex-needs families did not necessarily benefit from the program in the short term. When the research team presented the finding to the Program Managers' Committee, program managers stressed that lack of evidence of the program's short-term impacts on complex-needs families did not mean that the program could not help these families in the long term, which the research was not set up to evaluate.

The above feedback clearly fell beyond the study scope. It nevertheless caught the researchers' attention. After the meeting, the researchers realized that the finding that complex-needs families do not benefit from the program in evident ways could have significant implications for the future of the program. Among other possibilities, complex-needs families could be taken out of the program altogether. The research team was compelled to revisit the interview data. One sentiment that came out of interviews was that, for some complex-needs families, *KidsFirst* was the last lifeline in continually challenging family circumstances. Although this finding was hinted at in our report on preliminary findings, we decided to highlight it in the final report. In the research recommendations section, we also balanced the research evidence and the social responsibilities that the program managers tried to convey during the Program Managers' Committee meeting. We suggested that the program should provide specialized services to complex-needs families, which would take the workload off the paraprofessionals, who could then focus more on other families.

In retrospect, though the evaluation was designed to evaluate how the program worked to achieve the goals set up by the government, the community program workers also found a space where their interests as workers and as community members were respected and validated. To enable community programs to share their knowledge and to share ownership of the evaluation research, we learned that constant communication and space making, which involve both relationship building and power negotiation, are essential means that can make a difference.

Conclusion

We have reflected here on the research process of a participatory evaluation project that involved a tripartite partnership between researchers, community organizations, and government partners. We shared the researchers' working principle and strategies as well as the challenges faced and the lessons learned during the research process. We reiterated the importance of equal partnership in and joint ownership of a community-based research project, as extensively emphasized in the CBPR literature. In practice, we found regular communications and meetings, co-learning, and joint authorship most helpful in effecting and sustaining a collaborative and productive research relationship. The chapter adds to the existent CBPR literature by presenting the unique challenges that we encountered working in a tripartite relationship and by sharing our lessons in dealing with them. We showed that different standpoints of the research participants can lead to different voices and evidence that researchers need to reconcile. Furthermore, research participants might share more as the research relationship develops and as researchers create space that goes beyond the evaluation framework. We learned from our experiences that researchers need to treat contradictory data with care, fine-tune their report to reflect differing perspectives, and, most importantly, keep the lines of communication open even if they are in the position to make independent judgments. These lessons will be of practical use for other participatory evaluators who intend to produce evidence-based research findings that are ethically and socially responsible.

It is too early to tell what impact the evaluation research will have on the program given that the evaluation was concluded only six months ago. Nonetheless, this chapter has a few implications for other community-based research. First, once again, our study shows that, to build a successful collaborative relationship between academic researchers and non-academic research partners, it is important to value and mobilize the knowledge of the different partners in meaningful ways throughout the research process. Without the practical knowledge of the program managers and the policymakers, it would have been impossible for us to complete our study. Furthermore, without their understanding of the research process and their support, our field research would not have proceeded smoothly. Second, it is possible that, due to different standpoints and interests of the research partners, researchers can be presented with contradictory voices and evidence. In such cases, it is important to give due attention to all evidence, which might lead to new questions and issues that the study might or might not be designed to address. And third, we believe that it is important to keep the lines of communication open with all partners at all times. Constant communication can help to foster a safe space for people to articulate differences, to share, to dialogue, thereby strengthening a collaborative and productive research partnership.

Acknowledgements

The *KidsFirst* evaluation project would not have been possible without the involvement of a number of people. We acknowledge the insights, as well as the financial contributions, provided by the Early Childhood Development Unit (Gail Russell, Gary Shepherd, Rob Gates, Wendy Moellenbeck, and Murray Skulmoski) and each of the nine *KidsFirst* program sites. We would also like to thank the *KidsFirst* program managers, the staff at all sites, and all those who participated in the interviews and focus groups for providing the stories and experiences that formed the substance of the study. The study was developed with the guidance, support, and contributions of the many members of the *KidsFirst* Evaluation Team. It included the following *KidsFirst* investigators—Angela Bowen, Jody Glacken, Kathryn Green, Bonnie Jeffery, Thomas McIntosh, David Rosenbluth, and Nazmi Sari—as well as the following research staff—Darren Nickel and Fleur Macqueen Smith. Additionally, we recognize the work of Julia Hardy, Jillian Lunn, Karen Smith, Hayley Turnbull, Kathleen McMullin, and Taban Leggett in the process of data collection for the study.

References

Bandura, A. (1982). Self-efficacy mechanism in human agency. *American Psychologist, 37*(2), 122–147.

Botes, L., & van Rensburg, D. (2000). Community participation in development: Nine plagues and twelve commandments. *Community Development Journal, 35*(1), 41–58.

Bretherton, I. (1992). The origins of attachment theory: John Bowlby and Mary Ainsworth. *Developmental Psychology, 28*(5), 759–775.

Brown, L. D., & Vega, W. (1996). A protocol for community-based research. *American Journal of Preventive Medicine, 12*(4), 4–5.

Bubolz, M. M., & Sontag, M. S. (1993). Human ecology theory. In *Sourcebook of family theories and methods* (pp. 419–450). US: Springer.

Chávez, V., Duran, B., Baker, Q., Avila, M. M., & Wallerstein, N. (2008). The dance of race and privilege in CBPR. In M. Minkler & N. Wallerstein (Eds.), *Community based participatory research for health* (2nd ed.) (pp. 91–106). San Francisco: Jossey-Bass.

Cornwall, A., & Jewkes, R. (1995). What is participatory research? *Social Science and Medicine, 41*, 1667–1676.

Denzin, N. K., & Lincoln, Y. S. (Eds.). (2000). *Handbook of qualitative research* (2nd ed.). London: Sage Publications.

Gaventa, J. (1993). The powerful, the powerless, and the experts: Knowledge struggles in an information age. In P. Park, M. Brydon-Miller, B. L. Hall, & T.

Jackson (Eds.), *Voices of change: Participatory research in the United States and Canada* (pp. 21–40). Westport, CT: Bergin and Garvey.

Green, L. W., & Mercer, S. L. (2001). Can public health researchers and agencies reconcile the push from funding bodies and the pull from communities? *American Journal of Public Health, 91*, 1926–1929.

Green, L. W., George, M. A., Daniel, M., Frankish, C. J., Herbert, C. P., Bowie, W. R., et al. (1995). *Study of participatory research in health promotion.* Vancouver: University of British Columbia; Royal Society of Canada.

Hall, B. (1984). Research, commitment, and action: The role of participatory research. *International Review of Education/Revue internationale de l'education, 30*(3), 289–299.

Israel, B. A., Eng, E., Schulz, A. J., & Parker, E. A. (2005). *Methods in community-based participatory research for health.* San Francisco: Jossey-Bass.

Israel, B. A., Schulz, A. J., Parker, E. A., & Becker, A. B. (1998). Review of community-based research: Assessing partnership approaches to improve public health. *Annual Review of Public Health, 19*, 173–202.

Jones, L., & Wells, K. (2007). Strategies for academic and clinician engagement in community-partnered participatory research. *Journal of the American Medical Association, 297*, 407–410.

Kushner, S. (2005). Qualitative control: A review of the framework for assessing qualitative evaluation. *Evaluation, 11*(1), 111–122.

Letiecq, B. L., & Bailey, S. J. (2004). Evaluating from the outside: Conducting cross-cultural evaluation research on an American Indian reservation. *Evaluation Review, 28*(4), 342–357.

Minkler, M. (2000). Using participatory action research to build healthy communities. *Public Health Reports, 115*(2–3), 191–197.

Minkler, M., & Wallerstein, N. (2008). *Community-based participatory research for health from process to outcomes.* San Francisco: John Wiley and Sons.

Muhajarine, N., Macqueen Smith, F., Nickel, D., Russell, G., & the KidsFirst Evaluation Research Team & Collaborators. (2011). Using integrated KT to evaluate a complex early childhood intervention program. In J. Bacsu & F. Macqueen Smith (Eds.), *Innovations in knowledge translation: The SPHERU KT casebook* (pp. 32–36). Saskatoon: SPHERU. http://www.kidskan.ca/node/468.

Njoh, A. J. (2002). Barriers to community participation in development planning: Lessons from the Mutengene (Cameroon). *Community Development Journal, 37*(3), 233–248.

Patton, M. Q. (1997). *Utilization-focused evaluation: The new century text* (3rd ed.). Thousand Oaks, CA: Sage.

Patton, M. Q. (2002). *Qualitative research and evaluation methods.* Thousand Oaks, CA: Sage.

Puma, J., Bennett, L., Cutforth, N., Tombari, C., & Stein, P. (2009). A case study of a community-based participatory evaluation research (CBPER) project: Reflections on promising practices and shortcomings. *Michigan Journal of Community Service Learning,* (Spring), 34–47.

Reason, P., & Bradbury, H. (2006). *The handbook of action research: Concise paperback edition.* London: Sage.

Saskatchewan Education, Health, Intergovernmental, and Aboriginal Affairs and Social Services. (2002). *KidsFirst program manual.* Regina: Government of Saskatchewan.

Smith, P., & Bryan, K. (2005). Participatory evaluation: Navigating the emotions of partnerships. *Journal of Social Work Practice, 19*(2), 195–209.

Spingett, J., & Wallerstein, N. (2008). Issues in participatory evaluation. In M. Minkler & N. Wallerstein (Eds)., *Community-based participatory research for health from process to outcomes* (pp. 199–224). San Francisco: John Wiley and Sons.

Sullivan, M., Kone, A., Senturia, K. D., Chrisman, N. J., Ciske, S. J., & Krieger, J. W. (2001). Researcher and researched-community perspectives: Toward bridging the gap. *Health Education and Behavior, 28*(2), 130–149.

Vedung, E. (2004). *Public policy and program evaluation.* New Brunswick, NJ: Transaction Publisher.

Wallerstein, N. (1999). Power between evaluator and community: Research relationships within New Mexico's healthier communities. *Social Science and Medicine, 49,* 39–53.

Wallerstein, N., & Duran, B. (2006). Using community-based participatory research to address health disparities. *Health Promotion Practice, 7*(3), 312–323.

Williams, A., Labonte, R., Randall, J. E., & Muhajarine, N. (2005). Establishing and sustaining community-university partnerships: A case study of quality of life research. *Critical Public Health, 15*(3), 291–302.

STANDING BUFFALO FIRST NATION YOUTH
EXPLORING HEALTH AND WELL-BEING

Pammla Lusenga Petrucka, Roger Redman, Deanna Bickford, Sandra Bassendowski, Andrea Redman, Leanne Yuzicappi, Bev McBeth, Logan Bird, and Carrie Bourassa

Background

The community of Standing Buffalo First Nation, a small rural reserve in southern Saskatchewan, is the site of this project. The partnership with the community and a collaboration of academic researchers from the University of Saskatchewan, First Nations University of Canada, and University of Regina began with a shared vision to explore societal, cultural, and environmental influences impacting the "being" or "becoming well" of members of Standing Buffalo First Nation.

Currently, there is no specific ethics approval process within Aboriginal communities in Saskatchewan. Hence, all research was reviewed and approved for ethics by the participating institutions. The primary research ethics approval was obtained from the University of Saskatchewan's Behavioural Ethics Board.

In the initial phases of the partnership, a series of sharing circles was conducted to help identify the needs, aspirations, and preferred futures of specific groups (e.g., women, men, and youth) within the community. At this point of the research relationship, we were primarily exploring with the community and forming a more stable community partnership. Based on these findings and the aforementioned vision, the group met in a single facilitated session with representatives from the community (including Elders, chief, and council), universities, and other potential partners. The session was convened with a twofold purpose: to determine intent to work together on a research project, and, if this was established, to determine the direction and focus of such a project. The outcome of the session was a clear agreement to pursue an active community-based research (CBR) partnership with an initial project focusing on the health of youth in Standing Buffalo First Nation. Of significance, the Elders' guidance led to the titling of this research program as Ocanku Duta Amani, translated from the Dakota language to mean "paths to living well," with the inaugural program being entitled Paths to Living Well for On-Reserve Aboriginal Youth. It was determined by the partners that, although the research literature is replete with issues facing Aboriginal youth, ranging from disease incidence (e.g., diabetes, obesity), mental health challenges (e.g., suicide), and substance abuse issues (e.g., smoking, illicit drugs), the dominant biomedical model has proven unresponsive and unsuccessful in removing the burden of health experienced by Aboriginal youth. Researchers, policy-makers, and, more importantly, the communities are recognizing that it is impossible to talk about the health of Aboriginal children and youth without looking at the contributing historical, social, cultural, and environmental factors that impact the individual specifically and the community generally. Hence, the purpose of the Paths to Living Well research project was articulated as building understanding, identifying barriers and facilitators, and informing programs and policies related to the wellness challenges and opportunities of Aboriginal youth at Standing Buffalo First Nation.

The health of Aboriginal youth is both a local and an international issue. In this chapter, we describe a community-based research partnership that considers this paramount concern from the perspective of the youth themselves. Through visual methodologies, the research has considered being and becoming well from their perspective. We reflect on lessons learned in terms of the participants, the community, and the academic team. Broadly, the chapter considers the paths to living well as depicted by the youth in the form of visual products, including photographs, murals, and winter counts.

Research Process

During the facilitated session, the partners identified a preference for a visual method to be used in the research project since it was seen as culturally appropriate as well as attractive for the youth participants. The discussions led to a decision that photovoice is a potentially powerful and useful approach that would not only provide the youth with an opportunity to capture and convey what makes or keeps them well/unwell but also give them a life skill in using a camera. During the photovoice project, there have been between twelve and fourteen participants, with a core group of ten. The participants are all members of Standing Buffalo Dakota Nation and range from nine to eighteen years of age.

To date, the team has undertaken three sets of photovoice collection/dialogue sessions through which the youth have contributed over 700 pictures. Each youth was provided with a disposable camera and had access to a digital camera to secure the pictures. The participants individually selected three to five personal photographs for photo elicitation (Drew, Duncan, & Sawyer, 2010). These pictures were to represent best what they thought made them or kept them well/unwell. During these photo-elicitation sessions, participants shared their photographs and the stories related to them. Discussion was guided using the SHOwED format. It is a structured questioning technique and stands for the following:

- What do you **S**ee happening here?
- What is really **H**appening in this photo?
- How does this relate to **O**ur lives?
- Why do these issues **E**xist?
- What can we **D**o to address the issues?

Based on the guided discussions, the photographs were collectively arranged according to participant-generated themes. The three themes are described in the "outcomes" section of the chapter.

During this phase of the project, chief and council approached the research team about the possibility of using some of the messages to create a mural (see Figure 1) to hang in the powwow grounds. Standing Buffalo First Nation youth from within the project and beyond were invited to work with an artist in imagining, designing, and co-creating a series of four panels representing youths' paths to living well.

As the photovoice phase of the research drew to a close, the Elders and the community researcher advised the academic partner that the preferred next steps in the research should focus on traditional approaches to evoke further issues of being and becoming healthy for youth. This dialogue led to a suggestion of using winter counts, a traditional Dakota/Lakota historical calendar that portrays annual key community events as pictograms on an animal hide. On the advice of Elders, youth participants each create a personal winter count of the people/things that have contributed to their paths to living well. Each youth is given a hide and asked to provide one or more pictograms representing key contributing events/activities for each year of life. This phase of the research is currently under way, and at this point it is premature to draw any lessons or conclusions based on this approach.

Findings

As we considered the findings of this CBR, it was possible to categorize them in accordance with the three major partners: the participants, the community partners, and the academic partners.

The Participants

For the participants, the research has contributed to increased self-esteem and confidence. They have shown significant growth in their interactions with the Elders and community members. Their pride in the creation and presentation of their findings has contributed to their continued and consistent participation in the research activities. In consultation with community partners (including community meetings, contacts with the Elders, and meetings with chief and council), the following positive comments were noted regarding the youth.

Figure 1. Panel representing Standing Buffalo First Nation mural.

- Their involvement has increased their thinking about health; they are more aware of their roles as they relate to their own health as well as the roles that their environment and community play.
- Additionally, it has increased their thinking about their community—what its needs are and how they can lead and be part of the changes. ("I want to be a leader in the future . . . so I can make a difference 'cause there are still lots of problems here.")
- Both self-confidence and self-worth have increased due to their thoughts, opinions, and other contributions to the project being respected and valued.
- The youth felt a sense of belonging; they felt "included" in the group and that they belonged to something important.
- The youth had the opportunity to grow and become leaders of the next generation. As youth in the community, they took the time to join the group, were very committed to it, and took pride in their contributions. They have earned significant admiration from the community and team members for their dedication and hard work.
- In terms of learning about research, it was a positive experience for youth. We were surprised to learn that they had preconceived notions of "what" a researcher might be. After working together with us, the youth decided that we were "just like other regular people."
- Through their involvement, they had the opportunity to learn more about their culture and in turn pass on their knowledge to others. ("I did not know anything about winter counts, but now I can talk about them and feel really good about what I learned.")
- The youth were acknowledged for their contributions to the Paths to Living Well project. They received high school credits for their participation.

Through their photovoice analysis, the youth derived three major themes—*environment, cultural practices, people (across the generations)*—of what makes or keeps them healthy as on-reserve First Nations youth. All three themes were fairly equally represented in the photo-elicitation process.

Under the theme of environment, the youth included pictures of the sky, landscape, plants, and water. They showed both positive and negative situations, highlighting in their discussions the need to respect the environment, to understand humans' subservient role in attending to the environment, and the importance of maintaining balance between humans and nature. On the murals, this theme was similarly depicted with the hills, water, and sun.

Under the theme of cultural practices, the youth shared images of pow-wows, traditional costumes, beadings, tipis, artifacts, and hunting. They emphasized the need to honour the traditional practices; however, there was extensive discussion on the use of non-Dakota words to explain or "label" some of the events. In relation to this disjunction, the Elders contributed by working with the youth on Dakota language (since not all spoke Dakota) as they considered their experiences and perceptions. On the murals, this theme was reflected in drawings of tipis, sweetgrass, medicine wheels, and similar representative cultural items.

Under the theme of people (across the generations), the youth presented pictures of family members, friends, and role models within the community. Strikingly, they emphasized in the discussions the importance of connecting the very young to the very old—one participant even took a picture of a photograph of her grandfather and juxtaposed it with a newborn child. This sense of continuity and connectedness was also seen in the mural in community settings (e.g., a small village) and the unique combination of historical and modern depictions. Some of the youth were attracted to historical depictions of their village (tipis) and animals (traditional styles), while others were inclined to design pieces that represented how these things look today. For example, the finished mural contains a large (modern) eagle and wolf; it also portrays horses in a traditional (almost hieroglyphic) style and a tipi. Clearly, the youth see their environment as it is now, but they have strong respect for and keen awareness of what existed before and how their ancestors would have depicted it.

The Community Partners

For the community partners, the research project has built understanding of the research process generally and CBR specifically. The community has been given the opportunity to reflect on the products of the research, which has further contributed to an understanding of the perspectives, expectations, and needs of the youth participants. For example, the painted mural hung at the powwow ground portrayed factors leading to health and well-being for youth and was seen by hundreds of community members and visitors. This was seen as a powerful messaging tool. The murals served as starting points for discussions on health among community members and the youth as well as visitors attending the annual powwow and other events. Based on their shared cultural lens, they were able to derive meanings underlying the content of the murals for themselves and others that were vital to and reflective of their community. On a permanent basis, the mural panels are displayed in the community school library, where they continue to be viewed and discussed with the children and those passing through. We believe that the community has felt involved in the process and has been free to make suggestions, participate, and join at various times. In our opinion, this has taken away some of the mystique often attached to research.

There has been interest in the photovoice project, as it has come to be known in the community. People (non-team) have often joined sessions, especially during our summer cultural camp contributions, in order to participate in the opportunities. For example, in 2010, the CBR team decided to bring winter counts to all the summer camp participants, so each child was given a short orientation to winter counts by the Elder and youth research participants. Then all were provided with a small piece of hide and helped to create a small version of a winter count to take away (see Figure 2). Again, this allowed us to take our learnings and experiences beyond our core group. It is not clear at this point how pervasive the benefits might be across the wider community.

Figure 2. Sample winter count, summer 2010.

The Academic Partners

For the academic partners, the research journey has been revealing and challenging. Although community-based research was considered non-negotiable, it was interesting to consider the various interpretations of what this meant. For some, the community partners were the youth participants; for others, they were the research team (e.g., Elders, community researcher, participants, and academic researchers); yet others considered the entire Standing Buffalo First Nation as the community. It soon became apparent that these perspectives were not mutually exclusive but intersected and interfaced at various points during the process. For example, earlier in the research, there was more involvement of chief and council to establish the parameters and relationships, with reliance shifting more to the micro (i.e., individual youth participants) and meso (i.e., research team) level communities as we proceeded.

A key aspect of the meta-community (i.e., the entire community of Standing Buffalo First Nation) interaction involved community rituals and events. These political (elections), social (feasts/celebrations), and crisis (traumas/deaths) elements frequently impacted our research continuity. This was not equivalent to being flexible or adapting planning; it was more about developing a respectful and culturally appropriate relationship with the community as a whole.

Lessons Learned

A key lesson learned was that *engagement of the youth participants was best achieved through frequent activities to maintain interest and commitment.* We achieved this imperative by holding meetings on at least a monthly basis and "assigning" activities between sessions. Such activities included submitting digital pictures, depicting their individual "Indian names," and meeting with the community researcher to explore potential dissemination activities.

Additionally, two key learnings related to lines of communication and involvement. First, *the role of the Elders in the project was valued and valuable at all stages.*

- The Elders guided the project at inception by setting its name.
- The Elders participated in most sessions at which the youth were in attendance.

- The Elders provided a cultural experience at First Nations University of Canada re the tipi.
- The Elders are leading the winter count process, present at almost every session.
- The Elders guided the embodiment of community culture throughout the project.

Second, *the consistent support and involvement of chief and council were also integral to the team's productivity, transparency, and collaboration in the research process.* It became apparent in the latter stages of the process that there were missed opportunities to achieve and explicate the relationship fully. For example, the group expressed interest in terms of involvement in monitoring and funding the study tour, which did not occur. At times, it was difficult to know which levels of involvement and reporting were desired by chief and council as there were no documented roles and responsibilities decided at the outset. In future projects, this will take on a more focal role to enhance potential, and negotiation of roles will be determined at the outset of the project.

In terms of research approach, two key lessons were learned. First, *the community-based researcher's linkage to the participants (i.e., based out of the school) proved to be critical to coordination and project management.* Since the activities related to this project occurred outside school hours, it was important to have someone who could liaise with the participants between sessions as well as identify possible peak periods in school activity (e.g., sports tournaments, examination periods). Such events can (and in fact did) impact youth participation at times throughout the research.

Second, *research team flexibility and responsiveness (e.g., respectful approach) were essential in allowing the partnership to grow and foster innovation.* One example of this imperative was the move to winter counts as an alternative to the originally proposed digital presentation of the data. This suggestion arose from the community, and the academic research team embraced an opportunity to undertake a significantly challenging yet rewarding detour in the research process.

In addition, two general lessons were learned. First, *the community-based research and partnership processes had to be inclusive, transparent, active, and community specific to be effective and influential.* We experienced many "growing pains" and missed opportunities in our efforts only because we were all relatively new to this type of research. Through the early phases, there were challenges in learning the roles of community members and leaders, learning cultural ways of engaging, and building trust as a desired outcome rather than an expectation. At times, there were assumptions about processes and whether something was "important enough" to talk about. And, since we were working with a community that is by nature dynamic and amorphous, we often did not know of opportunities to share or work together unless someone identified them. The issue of being community specific was perhaps the most critical element since it allowed us to be responsive and respectful at a level that many other research processes might not have afforded us.

Second, *the visual research products (e.g., photographs, murals, and winter counts) are very appropriate and meaningful in the Aboriginal/First Nations context.* As we worked with the Elders, they fully endorsed and gave credence to the visual methodologies as being "good as it is the way the Dakota people shared" (personal conversation with Elder Ken Goodwill).

Next Steps

First, the project has yielded an emerging program influence related to youth activities being informed by these three thematic areas. For instance, we have worked with the local "culture camp" to ensure practical applications of each of the thematic areas (e.g., cultural practices: beading).

Second, the involvement with this population and this community has led to the foundational work on an activity initiative referred to as PL^2A^3Y (i.e., positive leadership, living, attitudes, and activities for Aboriginal youth). The program will be a culturally informed activity, play, and informal educational program that will complement existing programs on-reserve but build on the three thematic areas of being or becoming healthy as articulated in this research project.

Third, two graduate students are currently planning on continuing the work with Standing Buffalo First Nation, one in the area of PL^2A^3Y (master's level), the other in the area of emerging visual methodologies (Ph.D. level).

Conclusion

The experiences during our partnership have informed our understandings, expectations, and desired futures for community-based research. We have come to understand that CBR requires time, immersion, mutual respect, and shared vision. It also requires a shared vocabulary that is open and consistently applied. We remained open to including Dakota words, incorporated traditional terms, and sought guidance from the Elders, which often came in Dakota terms. Our expectations of CBR evolved during the course of our experience. For example, we came to recognize the expectations regarding communication and knowledge translation, including the respective roles of each member of the research team and the advisory roles of the community leaders and Elders, which were much more imperative than originally anticipated. We found that minor gaps or changes in either of these two dimensions could be time-consuming and often required revisiting. Finally, we are still articulating the desired futures for our CBR, but we recognize the need to increase the participation of all partners in all phases of the research, including development of the proposal, monitoring, evaluation, and, most importantly, knowledge translation/exchange. We are continuously seeking meaningful ways to achieve inclusive, culturally respectful approaches in these aspects.

Acknowledgements

Special thanks to our funders CIHR-IAPH (Canadian Institutes of Health Research-Institute of Aboriginal Peoples' Health), SPHERU, and the Saskatchewan Arts Board. There are no declared industrial links or affiliations for any of the partners. This chapter is dedicated to the Elders of Standing Buffalo First Nation whose wisdom and guidance have been invaluable—especially the late Ken Goodwill, who is sadly missed.

Articles of Interest

Castleden, H., Garvin, T., & Huu-ay-aht First Nation. (2008). Modifying photovoice for community-based participatory Indigenous research. *Social Science and Medicine, 66*(6), 1393–1405.

Christopher, S., Watts, V., Knows His Gun McCormick, A., & Young, S. (2008). Building and maintaining trust in a community-based participatory research partnership. *American Journal of Public Health, 98*(8), 1398–1406.

Cochran, P. A. L., Marshall, C. A., Garcia-Downing, C., Kendall, E., Cook, D., McCubbin, L., & Gover, M. S. (2008). Indigenous ways of knowing: Implications for participatory research and community. *American Journal of Public Health, 98*(1), 22–27.

Drew, S. E., Duncan, R. E., & Sawyer, S. M. (2010). Visual storytelling: A beneficial but challenging method for health research with young people. *Qualitative Health Research, 20*(12), 1677–1688.

Hergenrather, K. C., Rhodes, S. D., Cowan, B. G., & Pula, S. (2009). Photovoice as community-based participatory research: A qualitative review. *American Journal of Health Behavior, 33*(6), 686–698.

Poudrier, J., & Mac-Lean, R. T. (2009). "We've fallen into the cracks": Aboriginal women's experiences with breast cancer through photovoice. *Nursing Inquiry, 16*(4), 306–317.

KNOWLEDGE TRANSLATION STRATEGIES IN COMMUNITY-BASED RESEARCH
OUR DECISION-MAKER-BASED APPROACH

Fleur Macqueen Smith, Nazeem Muhajarine, and Sue Delanoy

Background and Context

This chapter describes the Understanding the Early Years (UEY) study in Saskatoon, a community-based research (CBR) project conducted as a community-university partnership. The study had a considerable impact on programs and policies for young children, their families, and their communities, due in part to the collaborative nature of its leadership. Its major approaches to knowledge translation were to conduct the study as a partnership between academic and community partners and to develop relationships with interested stakeholders over a long period of time, so that stakeholders could take ownership of the research process. To share this decision-maker-based approach, we have conceptualized it in a simple, five-step checklist that will guide others in conducting research in this way. The steps are (1) identify decision makers, (2) involve them early, (3) involve them often, (4) conduct research that they can use, and (5) give them results that they can understand. In this chapter, we describe our Healthy Children research program, one of our community-based research projects conducted as part of this program, and how we developed a checklist to conceptualize our approach to community-based research so as to maximize its impact. We conclude the chapter with a few thoughts on how to further community-based research.

Children are a good example of a population group whose compromised health has consequences not only for their life courses but also for society. Although older studies concluded that neighbourhood effects on children were very small after accounting for individual differences (Beauvais & Jenson, 2003), more recent studies have shown that, not only do neighbourhoods impact children's health outcomes independent of individual-level factors, but also a neighbourhood's social and physical characteristics can have even greater impacts on children's health outcomes than previously understood (Dunn, Rrohlich, Ross, Curtis, & Sanmartin, 2006; Muhajarine, Vu, & Labonte, 2006). Research on the impacts of physical and social environments on healthy childhood development is complex, and much of it suffers from a lack of conceptual clarity, a lack of data at the level of place of residence and activity that is meaningful to subjects, and little incorporation of social theories that explain the importance of context in the lives of individuals (Muhajarine et al., 2006). Children's health and development are the result of complex relationships among the child and a series of overlapping contexts such as the family, neighbourhood, region, and nation. Significant health inequities at the neighbourhood level have been found in the general population and for a range of child health and development measures in Saskatoon and beyond, such as immunization coverage, infant mortality, school readiness, physical activity and other health behaviours, mental health, and low birth weight; furthermore, these inequities are strongly associated with inequities in key social determinants of health, such as income, education, and Aboriginal cultural status (Avis, Tan, Anderson, Tan, & Muhajarine, 2007; Cushon, Vu, Janzen, & Muhajarine, 2011; Lemstra, Neudorf, & Opondo, 2006; Lemstra et al., 2007; Muhajarine & Vu, 2009; Muhajarine et al., 2006; Puchala, Vu, & Muhajarine 2010; Vu & Muhajarine 2010; Wright & Muhajarine, 2008).

To understand these health disparities better, and to address this research-policy-practice gap, in 1999 Nazeem Muhajarine launched a Healthy Children research program in SPHERU that explores the complex and dynamic interplay of factors shaping present and future health within the settings in which children are raised from conception to school age—the intrauterine environment, the family, neighbourhoods, and cities. We use diverse methods and designs, founded in epidemiology, but draw on a variety of social science disciplines such as education, sociology, and geography. We also engage community members, including policy-makers, throughout the research process. We believe that this methodological diversity and rigour in combination with community involvement comprise the greatest potential to produce scientifically valid results with high policy and practice impacts at different jurisdictional levels, capable of addressing health disparities.

The UEY National Initiative

In late 1999, as this program of research was being established, Muhajarine partnered with Sue Delanoy of Communities *for* Children, Saskatoon's planning council for a child- and youth-friendly community[1] to lead a seven-year study in Saskatoon called Understanding the Early Years. UEY was a national initiative funded by the Government of Canada to help community members understand the needs of children in their own communities so that they could better develop and deliver programs and services to meet these needs. Projects were community based, with funds held by community organizations that entered into a contribution agreement with the funder, the federal ministry of Human Resources and Skills Development Canada. These organizations reported in a number of ways to their wider communities on the health and well-being of children as they reached kindergarten. Working with school divisions, they administered the Early Development Instrument (EDI), a population-level tool from the Offord Centre for Child Studies completed by kindergarten teachers that measures children's "readiness to learn" in five areas: physical health and well-being, social competence, emotional maturity, language and cognitive development, and communication skills and general knowledge (Janus & Offord, 2007). They collected additional data on a subset of these children, using direct assessments and parent interviews, and local data to create a community inventory of programs and services for young children and their families. Project teams presented their findings by creating neighbourhood maps using geographic information systems (GIS) software to show data on children's EDI outcomes, access to community programs and services, and socio-economic characteristics of the neighbourhoods in which the children live based on census and local data. Projects also

•••••••••

1 This organization ceased operations in 2010; Delanoy has continued her community-based advocacy work for children and youth through several other organizations.

produced community action plans to act on these findings (Community Development and Partnerships Directorate, 2010).

Developing the Saskatoon Study as a Community-University Partnership

Understanding the Early Years projects were expected to be led by full-time community coordinators who served as intermediaries among the project's community coalition (a group of local people interested in early childhood development who provided an advisory role), the project researcher (typically engaged part time to develop a community inventory of programs and services and community mapping reports), and independent contractors engaged for other specific data collection (Community Development and Partnerships Directorate, 2010). However, the research, analysis, mapping, and reporting on results required for UEY projects were challenging for community-based organizations. When we were applying for a UEY project, we decided instead to share the leadership between a community partner and an academic partner and to bring in other people to contribute their skills as needed. For example, when first establishing the study's credibility with school divisions, we worked with local educational consultants who were former school administrators; later Delanoy hired a project coordinator at Communities *for* Children who spent half his time managing day-to-day activities for the UEY study, and Muhajarine added a full-time knowledge transfer specialist to his Healthy Children research team (Fleur Macqueen Smith) to assist with communication and knowledge translation.

Initially, the funder was uncertain about this high level of involvement by university-based researchers. The project clearly was intended to be community based, driven by community needs and expected to have an impact in that community. There might have been misgivings that the project could be "taken over" by an expert researcher who would then turn its focus from community impact to research process rigour. It is important to remember that this was more than a decade ago, when the concept of knowledge translation, focusing on improving the impacts of findings, was much less established in the research community. Although the academic team members initially were regarded with some suspicion when attending national meetings, over time the added value of partnering with well-established researchers who could provide ongoing guidance and leadership, rather than solely with more junior researchers who would be engaged for short time periods to handle specific tasks, became clear. Later a formative evaluation conducted for the funder demonstrated that researchers were key to individual project success (Evaluation Directorate Strategic Policy and Research Branch, 2009).

Using a Decision-Maker-Based Approach for CBR

During the course of the Saskatoon UEY project, we found that there was considerable interest in how we were conducting this community-based research and knowledge translation project. We used integrated knowledge translation, conducting the study as a partnership between academics and community members and nurturing relationships with interested stakeholders over a long period of time so that they could develop a sense of ownership of the research process. Partway through this project, we conceptualized how we worked by creating a practical checklist of this approach that drew on our own experiences in CBR and is further validated by the academic literature. It includes the five simple steps listed at the outset of this chapter. We have found this checklist, described in detail below, useful for our community-based and collaborative research projects.

Identify Decision Makers

Successful knowledge transfer is all about relationships (Canadian Institutes of Health Research, 2006). As anyone who has conducted research with non-academic partners knows, it takes a considerable amount of time and effort for these relationships to be fruitful. Identifying people who will be interested in collaborative research and committed to working together over a sustained period of time is a key first step. For our UEY study, decision makers included school administrators, teachers, parents, children's program and service providers, and government policy-makers at various levels. Partnering with Communities *for* Children enabled us to reach key constituencies in the community and beyond since it had built up credibility in the community and was seen as a convenor and connector of various organizations committed to children's well-being and community building.

When considering partners, researchers need to be sensitive to the differences in power between potential partners and take care not to exacerbate existing issues. These power differences are not just one way, such that university academics have power and community groups do not; power distribution is much more complex and subtle than that. There are multiple lines that divide people based on whom they know, what they know, and which resources and information they can access. There are also power differences between academics and between community members. We have to recognize these differences so that we can rebalance any power differentials in order to work together effectively.

Involve Them Early

Getting researchers and decision makers together early in the research process and keeping decision makers involved are critical in overcoming barriers. The findings from numerous studies have shown that the "best predictor of research use is the early and continued involvement of relevant decision-makers" (Denis & Lomas, 2003, p. 370). A large Canadian study of social scientists found that how researchers disseminated their findings was far more important to uptake than their research methods: the more resources researchers invested in connecting with decision makers, exchanging ideas, and disseminating their research findings, the greater their results were in terms of the use of their research (Landry, Lamari, & Amara, 2001).

Decision makers can help to drive the research process if they are involved from the outset. Relationships developed over time help to build the trust necessary for researchers and decision makers to work as true partners (Denis & Lomas, 2003). Our UEY project has had ongoing input from our research team, which has both community and academic members, and from our advisory group of policy-makers and program planners. We have found that it is worthwhile to take some time at the outset in the presence of all partners to agree on a shared vision and mission and to set clear expectations for roles and responsibilities. Although these guidelines do not prevent problems from arising, they do provide guidance to work through problems when they do arise.

Involve Them Often

Partnerships in which researchers and decision makers collaborate to improve the uptake of research results are widely recognized as effective (Walter, Davies, & Nutley, 2003). Additionally, both researchers and decision makers who have worked together have reported that the benefits of such collaborations outweigh the resource costs (Denis & Lomas, 2003; Ross, Lavis, Rodriguez, Woodside, & Denis, 2003). By working together earlier in the research process, both researchers and decision makers will have a better understanding of the views and expectations of the other community (Lomas, 2009), and they can move from "an understanding of knowledge as a product to an understanding of knowledge generation as a process" (Dickinson, 2004, p. 55). Lomas (2009, p. xiii) advises that the "multiple stages of the decision-making and research processes argue for far more ongoing communication of priorities, approaches, choice points and constraints between the two communities." Denis and Lomas (2003) point out that trust is the basis of collaborative research, and such trust can only build up over time and through frequent interactions and shared experiences.

These conceptualizations help illustrate how knowledge transfer can be an integral part of the research process, particularly through the use of partnerships and collaborative research processes. Lavis and colleagues wrote about this in their study examining the use of health services research in public policy making in Saskatchewan and Ontario: "Researchers (and research funders) should create more opportunities for interactions with the potential users of their research. They should consider such activities as part of the 'real' work of research, not a superfluous add-on" (2002, p. 146). Furthermore, research projects go through many stages, and going back to potential users of research findings is often important so that researchers do not lose touch with decision makers' needs. Although developing partnerships is more difficult and time-consuming, working together regularly fuels understanding of each other's needs and breaks down the walls between the "two communities" of research and policy/practice.

Our UEY research team was made up of both academic and community partners who worked regularly together, informed by an advisory group of policy makers and program planners. At each stage of the process, the team went back to the advisory group, comprised of potential users of the research findings, to make sure that the study was still reflective of their needs. This engagement not only informed the conduct of research but also helped to spread its findings further and more effectively than the team could have accomplished alone, thanks to the involvement of well-respected community advocates.

Conduct Research That They Can Use

CBR is effective because you start with the community and its needs—community questions drive the research cycle. For research to be relevant, we need to know which issues and questions decision makers have and how the research that we undertake together can help to address these issues.

In this way, we can achieve greater community buy-in and involvement, resulting in a much greater likelihood that research findings will be put into policy and practice. We have found that a higher level of community participation builds knowledge, experience, and community capacity for research. That said, as researchers we cannot compromise the rigour, standards, and relevance of the research that we conduct to appease decision makers—research needs to be methodologically sound as well as relevant to the world of policy and practice to have the best chance of making an impact (Macqueen Smith, Nickel, Shan, & Muhajarine, 2011). Without credible results, there is no application of research in practice or policy.

Give Them Results that They Understand

Academics are used to conducting research and publishing their findings in academic journals and presenting at academic conferences. However, decision makers do not usually read academic journals or attend research conferences, so we need to present our findings in ways that are more widely accessible. For the Saskatoon UEY study, we disseminated our findings through colourful fact sheets, plain-language research reports, and newsletter articles, in print and online. We made extensive use of GIS mapping, showing our results by neighbourhoods on colourful Saskatoon maps that made it easy to identify trends. We also made numerous presentations, organized day-long community forums, and met regularly with the early years coalition of stakeholders. We found that people responded very positively to these plain-language messages, which shared our findings in a clear, compelling manner, effectively answering the classic "So what?" question that decision makers often raise. As the study results were disseminated, we also found that they enhanced our community partners' reputations as those who both conduct research and use research to inform their own advocacy and program development work.

Key Findings and Impacts of the Saskatoon Study

As part of the Saskatoon study, in 2001, 2003, and 2005 kindergarten teachers evaluated their students using Offord's Early Development Instrument. The UEY study team also conducted a community survey of all programs for kindergarten children and their families and a neighbourhood observation survey of all fifty-six residential neighbourhoods at that time in Saskatoon. Direct assessment of developmental indicators in children and interviews with their parents were done by Statistics Canada, with a random selection of 500 kindergarten children and parents in 2001 and 2005, as part of the National Longitudinal Survey of Children and Youth Community Study (KSI Research International Inc., 2002). Analysis of the findings showed that Saskatoon children as a whole were lagging behind the national norms in three of the five areas measured: physical health and well-being, language and cognitive development, and communication skills and general knowledge. Saskatoon parents also had lower scores on positive parenting skills, and Saskatoon mothers had lower scores on maternal mental health than national norms. Both factors might have contributed to higher levels of behavioural problems that we identified in Saskatoon children (1.5 times the national norms in 2001). Furthermore, this was of concern since Saskatoon also had more low-income families and single-parent families than the national average—families that could face more parenting difficulties as a result of their socio-economic circumstances.

To share these findings, we produced two community mapping reports (Muhajarine, Delanoy, Hartsook, & Hartsook, 2003; Muhajarine et al., 2005) and a series of focused fact sheets on literacy, maternal mental health, children's behavioural issues, and positive parenting, with findings from the study and tips for parents and caregivers on addressing issues (available on our knowledge dissemination website, kidsKAN, www.kidskan.ca [select UEY from the projects menu or go to kidskan.ca/uey]). We also made dozens of presentations to interested stakeholders throughout the course of the study.

We have been able to identify many instances in which our decision-maker-based approach made it possible for stakeholders to use our findings in policy and program development for young children in Saskatchewan. As the study was concluding, we held a community forum at which we invited key stakeholders to share the study's impacts on their work. Both the Saskatoon public and separate school boards reported that they had based major initiatives, such as literacy programs and literacy-enhanced, full-day, every-day kindergarten programs, on the Saskatoon UEY research. The Saskatoon Public Library told us that it had improved access to services where the study showed that people were underserviced, initially through its book trailer program and subsequently by opening a new branch on 20th Street after conducting its own study. Earlier we had learned that the provincial government, using in part our findings, had funded additional speech and language pathologists to work with children. And in 2009, after EDI data had been collected in various parts of the province through UEY projects in the past decade, the provincial government began collecting EDI data on all children across the province to monitor their outcomes over time and shared these data with us.

This project has facilitated many other CBR projects on early childhood development. Muhajarine and the Healthy Children research team partnered with three other Understanding the Early Years projects to provide research and community mapping expertise: Northeast Saskatchewan (2005–08), Southeast Saskatchewan (2007–10), and Moose Jaw-South Central Saskatchewan (2007–10). In the Saskatoon project, initially we had to work diligently to establish our credibility with the school boards to have their teachers report on kindergarten children using the EDI, so we worked with local educational consultants who were former school administrators. However, in these subsequent projects, our community partners reported that they had no trouble getting their school boards to participate, based on the observed benefits that the Saskatoon boards reaped.

In 2006, we were contracted by the Ministry of Education to evaluate full-time kindergarten pilot programs that several school boards were running (Muhajarine, Evitts, Horn, Glacken, & Pushor, 2007). In 2007, we partnered with the provincial government to conduct a three-year evaluation of *KidsFirst*, its early childhood development intervention program for very vulnerable young children and their families (Muhajarine, Nickel, Shan, & *KidsFirst* Evaluation Team, 2010). Also in 2007, we launched a provincial community of practice, kidsKAN, the Saskatchewan Knowledge to Action Network for early childhood development (www.kidskan.ca), which received Knowledge to Action funding from the Canadian Institutes of Health Research in late 2008. These subsequent projects were all conducted as community-based projects using the same kind of decision-maker-based approach since we had found it so effective. The more we work this way, the more we like working this way; although it is more challenging, it is also much more likely to have positive impacts on policy and practice, our ultimate goal.

Why Was This Approach Successful, and How Can We Replicate It?

We believe that several factors helped to make this partnership so fruitful. It got off to a strong start when we collaborated to write the proposal, and once it was funded we were careful to undertake the joint leadership that had been proposed. We learned from each other about our respective work cultures, the academic world, and the non-governmental, community-based organizations so that we could move more easily among them. Our decision-maker-based approach to working together gave stakeholders many opportunities to embrace the project and implement its findings. And, finally, we worked from the understanding that the UEY reports and fact sheets that we produced would serve as "talking points" to help illustrate and drive change, not as recipes to "fix" problems in our community. People sometimes put great faith in research reports and then experience considerable disappointment when the findings are not readily translated into practice. As has been noted in the knowledge translation literature, research evidence is only one ingredient of how policy and practice decisions are made (Lomas, 1997).

As to how we can sustain ongoing partnerships, this question can be considered from several angles: the university, the community, and funding agencies. Within the university, we need to continue to conduct research with people from different disciplines. When you broaden the base of stakeholders, you can broaden the base of interest and capacity. We also need to continue our work to establish CBR as a legitimate way of conducting research. In the traditional academic merit system, there is still often greater reward for publishing in a peer-reviewed journal with high-impact value, even if the article has little to no impact on policy and practice, than for publishing a policy paper that goes to government and has a significant impact on policy. That is why it is so important to make sure that CBR is highly valid as well as highly relevant. We also believe that community groups need to own the research process more by helping to set the research agenda, not just handing over data to an academic or signing a letter of support for a research project. Community agencies that are delivering programs should consider setting aside some resources for evaluation so that they can review and improve their programs. Finally, we hope that funders will recognize community-university partnership research as a standard way of doing research, lasting more than one budget cycle or one presidency of a funding agency. We need funders to really understand how these types of research work so that they can better facilitate them in various ways, such as recognizing the desirability of community and academic co-principal applicants for projects and understanding that effective knowledge translation requires considerably more time and funding than traditional researcher-driven research.

In CBR, we all come together—academics, students, community partners, and other stakeholders—to try to find answers to researchable questions. Although working this way is much harder than conducting researcher-driven studies, it can also be more energizing and much more likely to have impacts on policy and practice. At stake is more than research, more than writing a report or publishing a paper. It's about knowing that our research is making a difference in the wider world and being able to see the results. Most importantly, it is about helping to make our communities better places for everyone living in them.

Acknowledgements

We acknowledge the following funders of the research and knowledge translation projects mentioned above: the Government of Canada (Understanding the Early Years studies); the Government of Saskatchewan (full-time kindergarten and *KidsFirst* evaluations); the Canadian Population Health Initiative-Canadian Institute for Health Information, MITACS-Accelerate, and the College of Medicine at the University of Saskatchewan (*KidsFirst* evaluation); and the Canadian Institutes of Health Research (kidsKAN). We also acknowledge the assistance of Jeffrey Smith, Ph.D., in developing the checklist.

References

Avis, K., Tan, L., Anderson, C., Tan, B., & Muhajarine, N. (2007). Taking a closer look: An examination of measles, mumps, and rubella uptake in Saskatoon. *Canadian Journal of Public Health, 98*(5), 417–421.

Beauvais, C., & Jenson, J. (2003). *The well-being of children: Are there "neighbourhood effects"?* Ottawa: Canadian Policy Research Networks. http://www.cprn.org/.

Canadian Institutes of Health Research. (2006). Lessons learned. In *Evidence in action, acting on evidence: A casebook of health services and policy research knowledge translation stories.* Ottawa: Canadian Institutes of Health Research.

Community Development and Partnerships Directorate, Human Resources and Skills Development Canada. (2010). *Overview of the Understanding the Early Years Initiative.* http://www.hrsdc.gc.ca/.

Cushon, J., Vu, L. T. H., Janzen, B., & Muhajarine, N. (2011). Neighborhood poverty impacts children's physical health and well-being over time: Evidence from the Early Development Instrument. *Early Education and Development, 22*(2), 183–205.

Denis, J. L., & Lomas, J. (2003). Convergent evolution: The academic and policy roots of collaborative research. *Journal of Health Services Research and Policy, 8*(Suppl. 2), 1–6.

Dickinson, H. (2004). A sociological perspective on the transfer and utilization of social scientific knowledge for policy-making. In L. Lemieux-Charles & F. Champagne (Eds.), *Using knowledge and evidence in health care: Multidisciplinary perspectives* (pp. 41–69). Toronto: University of Toronto Press.

Dunn, J. R., Rrohlich, K. L., Ross, N., Curtis, L. J., & Sanmartin, C. (2006). Role of geography in inequalities in health and human development. In J. Heymann, C. Hertzman, M. L. Barer, & R. G. Evans (Eds.), *Healthier societies: From analysis to action* (pp. 237–263). New York: Oxford University Press.

Evaluation Directorate Strategic Policy and Research Branch, Human Resources and Skills Development Canada. (2009). *Formative evaluation of the Understanding the Early Years Initiative—June 2009*. Ottawa: Government of Canada. http://www.hrsdc.gc.ca/.

Janus, M., & Offord, D. (2007). *Early Development Instrument: A population-based measure for communities (EDI)*. http://www.offordcentre.com/readiness/.

KSI Research International Inc. (2002). *Understanding the early years: Early childhood development in Saskatoon, Saskatchewan*. Ottawa: Applied Research Branch, Strategic Policy, Human Resources Development Canada. http://www.kidskan.ca/.

Landry, R., Lamari, M., & Amara, N. (2001). Utilization of social science research knowledge in Canada. *Research Policy, 30*, 333–349.

Lavis, J., Ross, S., Hurley, J., Hohenadel, J., Stoddart, G., Woodward, C., et al. (2002). Examining the role of health services research in public policymaking. *Milbank Quarterly, 80*, 125–154.

Lemstra, M., Neudorf, C., & Opondo, J. (2006). Health disparity by neighbourhood income. *Canadian Journal of Public Health, 97*(6), 435–439.

Lemstra, M., Neudorf, C., Opondo, J., Toye, J., Kurji, A., Kunst, A., & Tournier, C. (2007). Disparity in childhood immunizations. *Pediatrics and Child Health, 12*, 847–852.

Lomas, J. (1997). *Improving research dissemination and uptake in the health sector: Beyond the sound of one hand clapping*. Policy Commentary C97-1. Hamilton: McMaster University Centre for Health Economics and Policy Analysis.

Lomas, J. (2009). Improving research dissemination and uptake in the health sector: Beyond the sound of one hand clapping [introduction]. In S. Straus, J. Tetroe, & I. Graham (Eds.), *Knowledge translation in health care: Moving from evidence to practice* (pp. 2–35). Oxford: BMJ Books/Wiley-Blackwell.

Macqueen Smith, F., Nickel, D., Shan, H., & Muhajarine, N. (2011). Early childhood intervention in the community makes sense, but does it really work? Findings from our three-year collaborative study. In *Population health intervention research casebook* (pp. 41–44). Ottawa: Canadian Institutes of Health Research—Canadian Population Health Initiative—CIHI. http://www.cihr-irsc.gc.ca/.

Muhajarine, N., Delanoy, S., Hartsook, B., & Hartsook, L. (2003). *Community mapping for children in Saskatoon (2001 data)*. Saskatoon: SPHERU. http://www.kidskan.ca/.

Muhajarine, N., Evitts, T., Horn, M., Glacken, J., & Pushor, D. (2007). *Full-time kindergarten in Saskatchewan part two: An evaluation of full-time kindergarten programs in three school divisions*. Saskatoon: CUISR.

Muhajarine, N., Nickel, D., Shan, H., & *KidsFirst* Evaluation Team. (2010). *Saskatchewan KidsFirst program evaluation: Summary of findings and recommendations*. Saskatoon: SPHERU. http://www.kidskan.ca/.

Muhajarine, N., & Vu, L. (2009). Neighbourhood contexts and low birth weight: Social disconnection heightens single parents' risks in Saskatoon. *Canadian Journal of Public Health, 100*(2), 130–134.

Muhajarine, N., Vu, L., Dyck, C., Macqueen Smith, F., Delanoy, S., & Ellis, J. (2005). *Community mapping for children in Saskatoon (2003 data)*. Saskatoon: SPHERU. http://www.kidskan.ca/.

Muhajarine, N., Vu, L., & Labonte, R. (2006). Social contexts and children's health outcomes: Researching across the boundaries. *Critical Public Health, 16*, 205–218.

National Collaborating Centres for Public Health Knowledge Translation Awards 2011. (2011). http://www.nccph.ca/.

Puchala, C., Vu, L., & Muhajarine, N. (2010). Neighbourhood ethnic diversity buffers school readiness impact in ESL children. *Canadian Journal of Public Health, 101*(9), Suppl. 3, S13–S18.

Ross, S., Lavis, J., Rodriguez, C., Woodside, J., & Denis, J. L. (2003). Partnership experiences: Involving decision-makers in the research process. *Journal of Health Services Research and Policy, 8*(Suppl. 2), 26–34.

Vu, L., & Muhajarine, N. (2010). Neighbourhood effects on hospitalization in early childhood. *Canadian Journal of Public Health, 101*(2), 119–123.

Walter, I., Davies, H., & Nutley, S. (2003). Increasing research impact through partnerships: Evidence from outside health care. *Journal of Health Services Research and Policy, 8*(Suppl. 2), 58–61.

Wright, J., & Muhajarine, N. (2008). Respiratory illness in Saskatoon infants: The impact of housing and neighbourhood characteristics. *Social Indicators Research, 85*(1), 81–95.

CONCLUSION
Isobel M. Findlay and Diane Martz

Most of the chapters in this book have been written in collaboration by community and academic partners, illustrating the strength of the relationships that have been forged through the research projects reported here. Every chapter in the book talks about the need to build and sustain trusting relationships, which are a key factor in the success and sustainability of a community-engaged research project. Relationships of trust determine who will be included in and excluded from full participation in projects (DeSantis) and whether hidden transcripts will become public discourses (Shan et al.).

Common Themes
The chapters in this book illustrate the diversity of community-engaged research, with projects falling along a continuum from those driven by a community that strictly controlled academic, government, and NGO participation (DeSantis) to those initiated and led by academics (Macqueen Smith et al.; Shan et al.). In the middle of this continuum are projects that appear to be closer to the CBR ideal of equal partnerships (Petrucka et al.) and those that have evolved over time to resemble equal partnerships more closely (Ebbesen et al.; Macqueen Smith et al.). The question of "Who is the

community?" in these community-engaged research projects is answered in many different ways from small rural Aboriginal communities (Petrucka et al.), to the community of northern trappers and anyone else who was interested (Findlay et al.), to a number of examples of communities of interest consisting of academics and government departments (Ebbesen et al.) or academics, government departments, and local organizations (Macqueen Smith et al.; Shan et al.). Some projects defined the members of their communities very restrictively (DeSantis), while others allowed their definitions to become very broad and inclusive (Findlay et al.; Petrucka et al.).

Time is a pervasive theme in these chapters: time to build trust, time to think through the issues, time to understand, time to do community-engaged research well and ethically. Both community and academic researchers are pressed to find the time that it takes for community-engaged research given their busy schedules, academic researchers need to present timely results to the community, and they need to take time to talk through issues, educate students in community-engaged research, and put findings into action. Time is also inherent in the reflexivity suggested by Findlay et al., Chopin et al., and Ebbesen et al. as a cornerstone of their approach to community-engaged research. As many of the chapters illustrate, reflexivity must be combined with effective communication among partners as well as willingness and flexibility to apply the lessons learned to future practice.

Many authors report on the fascinating challenge of brokering the interests of various partners in research. DeSantis talks about an "inclusion/exclusion dynamic," Diamantopoulos and Usiskin suggest that, for their project, the word community suggests a false unity, although they also recognize that community itself "is in a continuous state of contention and development," and Ebbesen et al. recount the ebb and flow of interests aligning at times and misaligning at other times. Ebbesen et al. also discuss the difficulty with some partners with the advocacy inherent in community-based participatory action research when it conflicts with their agencies' interests. Nevertheless, the projects described throughout the book have produced advocacy and change.

Challenges to the Practice of Community-Engaged Research

The chapters in this book celebrate many achievements and lessons learned, the rich rewards of friendships made, and mutual commitments to community and academic development on the ten-year, community-based research journeys of SPHERU and CUISR, but they also identify ongoing opportunities and challenges. Given the contradictory effects of globalization and neoliberalism, partnering is the new norm in research as in other contexts. Community-based organizations needing to support funding applications and policy changes with timely, relevant research are encouraged to form partnerships with universities (Stoecker, 2007). Yet the complex politics and cultures of productive partnerships are often underestimated (Macdonald & Chrisp, 2005), and many are only now beginning to come to terms with some of the ethical implications of community-university partnerships in current funding climates (Stewart, 2011a, 2011b). Similarly, Cornwall and Brock (2005) warn that buzzwords such as *participation* and *empowerment* can be pretexts for development that legitimates intervention, aggravates domination, and entrenches business as usual. Even for SPHERU and CUISR, founded on participatory principles and action-oriented research, building equitable partnerships and meaningful participation for change in the face of systemic and other issues remains as challenging as it can be enriching for all involved.

In line with this larger partnering phenomenon, SSHRC's Community-University Research Alliances (CURAS), for instance, bring together communities and universities to build knowledge on issues facing society. Yet, like other mainstream institutions, SSHRC can be slow to change its protocols to accommodate partner needs and to credit partner contributions. Despite its emphasis on partner equality, SSHRC depends on already overstretched community-based organizations contributing in-kind, personnel, and/or financial resources, while being warned to avoid the "formulaic" in letters confirming willingness "to complete activities assigned to it." Although SSHRC offers salary replacement stipends covering up to fifty percent of the cost of replacing staff research investigators, the community typically commits disproportionate resources to research projects without appropriate recognition. Although both SSHRC and the Canadian Institutes of Health Research (CIHR) still locate funding controls in academic institutions, promising initiatives at the University of Saskatchewan (piloted by CUISR and guided into policy by the Research Ethics Office) and the University of Regina (see the chapter by Kubik et al.) are facilitating payment of honoraria under $100 to research participants who will not be made vulnerable by being required to provide identification and Social Insurance Numbers (University of Saskatchewan, 2011a).

The chapter by Martz and Bacsu importantly reflects both university and community perspectives on the negotiation of ethical research relationships and takes us beyond traditional concerns with avoiding harm and maintaining confidentiality and anonymity to questions about benefits and control of the purse strings (see too the chapters by DeSantis and Findlay et al.). Similarly, the chapter by Kubik et al. enhances understanding of the complex ethical challenges of safety and confidentiality facing researchers working with vulnerable populations.

Still, many community groups and organizations continue to feel over-researched while failing to see sufficient benefits to justify the risks of participating in research that might, for example, result in reduced funding (see Diamantopoulos & Usiskin and Shan et al.). Although some projects can claim remarkable impacts on policy and programming (Macqueen Smith et al.), others have extended timelines that can leave community partners skeptical about returns on their investments. Similarly, community contributions are "often forgotten" after projects are completed, and communities do not always enjoy a legacy of ongoing "networks, resources and skills development" that might support them in continuing research on their own (Tang, 2008, pp. 242–243). On the other hand, this book offers a number of examples of community-university relationships that have endured for a number of years through a succession of linked projects that produce positive impacts for the partners involved (Ebbesen et al.; Macqueen Smith et al.; Petrucka et al.).

Although the community might have unrealistic expectations of what can be achieved in one project, it is often content to have stories heard, new options made visible, and its youth engaged and empowered to participate in traditional activities (see the chapters by Findlay et al. and Petrucka et al.).

As an "engaged university," the University of Saskatchewan likewise invests in "focused and collaborative endeavours...where research embraces critical issues of importance to society...and where partnerships...make the university's contributions visible and meaningful" (University of Saskatchewan, 2011b). Like the University of Regina, which highlights CBR as part of its signature theme, knowledge creation and discovery (University of Regina, 2011), the University of Saskatchewan is now explicitly committed to CBR in institutional planning and priorities. But such recognition can be a mixed blessing if institutional cultures and value systems do not change. Like many others, our own universities struggle to live up to CBR commitments.

Although CBR is on universities' radar, it—together with its public outreach, co-authorship, and multi-mediated outputs—often remains suspect among our academic peers for its accessible style and activist orientation. The rigour of CBR overlooked and often insufficiently discussed and defended by CBR researchers (Roche, 2008), gatekeepers in peer review processes associated with granting councils and journals continue to police what counts as authoritative knowledge and practice, often discounting it as "advocacy" tainted by application rather than "real scholarship" while remaining blind to the interests served by mainstream scholarship, whose assumptions of gender, race, and class can be so natural as to be invisible. And a focus on "A"-list journals often excludes the venues for engaged CBR and/or requires the recentring of mainstream thinking that perpetuates the sorts of inequalities—the "relations of ruling" (Smith, 1990)—that participatory action research is designed to unsettle and dislodge. CBR remains something of a poor cousin to the "disinterested" scholarship of quantitative research for refereed academic journals. Practising CBR can be "a form of professional suicide" (Ibáñez-Carrasco & Riaño-Alcalá, 2011, p. 81).

To attempt to address these issues (also articulated in Martz & Bacsu and Macqueen Smith et al.), a recent initiative led by Community Campus Partnerships for Health from the University of Guelph, including the University of Saskatchewan and the University of Regina, is exploring how university policy and practice need to change to develop and reward community-engaged scholarship more adequately.

Although SSHRC and universities laud interdisciplinarity, these discourses continue to produce hierarchies of knowledge and institutional practices that still do not always know how to value interdisciplinary work. Respect for experiential knowledge and openness to different knowledge ecologies, especially Indigenous knowledge, remain uneven. Similarly, institutional supports and sanctions remain unevenly applied, and community and university expectations of appropriate outcomes are not always well aligned (see the chapter by Ebbesen et al.). SSHRC invites diverse dissemination and qualitative measures of research impacts, but the standard curriculum vitae remains a numbers game, with refereed articles the priority and community relationships, training, and presentations relegated at best to community contributions at the end.

Student training remains a priority (see the chapters by Chopin et al. and Findlay et al.), and there has been progress in the increasing availability of courses in qualitative methods and even across the range of CBR methods, with opportunities for internships, practicum experience, and community service learning (see Chopin et al.). If many disciplines remain in denial about the role of research and education in past and ongoing colonial incursions (Smith, 1999), a greater focus on diverse knowledge mobilization activities (see the chapter by Petrucka et al.) adds to understanding the value of CBR. It also adds to the numbers of those willing to engage in research that is participatory and decolonizing in approach. A new generation educated in anti-colonial or anti-oppressive strategies is redrawing community-university relationships (see Kubik et al.), bringing "hidden transcripts" into public discourse (see Shan et al.), and clarifying what we all have to gain from new meaning making for self-empowering and creative communities.

This book marks a decade of achievements and lessons learned, the rich rewards of friendships made, and mutual commitments to community and academic development on the community-based research journeys of SPHERU and CUISR. The chapters illustrate strong productive research relationships forged over time, many of which are still ongoing and are based on genuine inclusion of community knowledge and voice. There are many examples where diverse knowledges were brought together, when community partners with in-depth practical knowledge complement academic knowledge (Kubik et al. and Macqueen Smith et al.) and traditional knowledge provides direction to both the research project and to the communities involved (see for example Petrucka et al. and Findlay et al.) Petrucka et al. and Kubik et al., among others, provide marvelous examples of methodological innovation through the use of new tools to depict community issues and new groups to include as research participants. Ebbesen et al., Findlay et al., and Shan et al. provide us with examples of academically rigorous research accomplished in the context of CBR. The chapters provide numerous examples of community engaged research with significant impacts through changes in policy, processes, and programming (DeSantis, Ebbesen et al. and Macqueen Smith et al.); changing perspectives on the appropriate mandate of organizations (Ebbesen et al.); community empowerment (Findlay et al. and Petrucka et al.); and personal transformation (for example Martz & Bacsu, Findlay et al., and Chopin et al.)

Finally, the academic and community researchers who collaborated to write the chapters for this book have all illustrated that their research journey unfolds on a foundation of the ethical principles that have defined the relationships among research teams members and their participants and enabled the research teams to respond effectively and ethically to arising issues.

References

Cornwall, A., & Brock, K. (2005). What do buzzwords do for development policy? A critical look at 'participation,' 'empowerment,' and 'poverty reduction.' *Third World Quarterly, 26*(7), 1043–1060.

Ibáñez-Carrasco, F., & Riaño-Alcalá, P. (2011). Organizing community-based research knowledge between universities and communities: Lessons learned. *Community Development Journal, 46*(1), 72–88.

Macdonald, S., & Chrisp, T. (2005). Acknowledging the purpose of partnership. *Journal of Business Ethics, 59*(4), 307–317.

Roche, B. (2008). *New directions in community-based research.* Toronto: Wellesley Institute.

Smith, D. E. (1990). *Texts, facts, and femininity: Exploring the relations of ruling.* London: Routledge.

Smith, L. (1999). *Decolonizing methodologies: Research and Indigenous peoples.* London: Zed Books.

Stewart, P. (2011a). President's column: National standards needed for research integrity. *CAUT Bulletin*, March, A3.

Stewart, P. (2011b). President's column: Collaborations: Are universities sacrificing integrity? *CAUT Bulletin*, May, A3.

Stoecker, R. (2007). The research practices and needs of nonprofit organizations in an urban center. *Journal of Sociology and Social Welfare, 34*(1), 97–119.

Tang, S. S.-T. (2008). Community-centred research as knowledge/capacity building in immigrant and refugee communities. In C. R. Hale (Ed.), *Engaging contradictions: Theory, politics, and methods of activist scholarship* (pp. 237–262). Berkeley: University of California Press.

University of Regina. (2011). *Working together towards common goals: Serving through research.* Strategic Research Plan 2010–2015. Regina: University of Regina.

University of Saskatchewan. (2011a). *Guidelines and procedures: Completing a research participants funding requisition.* http://www.usask.ca/.

University of Saskatchewan. (2011b). *Institutional planning and assessment. 2nd integrated plan: Frequently asked questions.* http://www.usask.ca/.

CONTRIBUTORS

Juanita Bacsu, M.A., is working on her Ph.D. in Community Health and Epidemiology at the University of Saskatchewan. She has worked as a research officer at SPHERU since 2009. Her research interests include support systems and seniors' health, rural health and public policy, and knowledge translation.

Sandra Bassendowski, Ph.D., is a professor with the College of Nursing, University of Saskatchewan. Her research focuses on the integration of technology in teaching and learning environments and the use of mobile devices to support nursing practice. She encourages students to use social networking tools for engagement and participation in the co-creation of course content.

Maria Basualdo, M.A., Ph.D. (candidate), is CUISR's former community-university research liaison. She was the face of CUISR for many community-based organizations, organizing new research partnerships and opportunities, doing research in the field, designing and delivering workshops and presentations, and motivating and mentoring researchers.

Deanna Bickford is a master's student at the College of Nursing, University of Saskatchewan. She has worked as a research assistant at Standing Buffalo First Nation.

Logan Bird is an undergraduate nursing student at the University of Saskatchewan. He worked as an Indigenous Peoples' Health Research Centre funded summer student at Standing Buffalo First Nation.

Carrie Bourassa, Ph.D., is an associate professor of Indigenous health studies at First Nations University of Canada. She completed her Ph.D. (social studies) in 2008. She is also the nominated principal investigator of the Indigenous Peoples' Health Research Centre. Carrie is Métis and belongs to the Regina Riel Métis Council.

Nicola Chopin, B.Sc., M.A., a former strategic research coordinator with CUISR, began working with CUISR as a practicum student in 2007. She was hired as CUISR's evaluation coordinator in August 2008 and moved into the position of strategic research coordinator in April 2009. She was responsible for writing proposals, managing research projects, and mentoring students.

Louise Clarke, Ph.D., associate professor, was involved with CUISR since 2001 and university co-director from 2006 to 2011. She also taught industrial relations and organizational behaviour in the Edwards School of Business and was a scholar of the Centre for the Study of Cooperatives, all at the University of Saskatchewan. She retired from the university in June 2011 but retains her research interests generally in social justice and specifically in community economic development, community-university partnerships/CBR, and labour-management relations.

Sue Delanoy has been widely recognized for her twenty-five years of work as a community-based advocate for children and youth. At present, she is the executive director of the Elizabeth Fry Society of Saskatchewan.

Gloria DeSantis, Ph.D., is a research associate at SPHERU and a lecturer in public policy, health, and interdisciplinary studies, University of Regina. Her research focuses on health, social justice, and the role of non-profit social service organizations in facilitating healthier communities. She has twenty years of experience working on CBR initiatives given her employment in the non-profit sector.

Mitch Diamantopoulos, Ph.D., is a founder of Saskatchewan's two independent city papers, *Prairie Dog* magazine in Regina and *Planet S* magazine in Saskatoon. He now heads the School of Journalism at the University of Regina and is a centre scholar with the University of Saskatchewan's Centre for the Study of Cooperatives.

Lori Ebbesen, Ed.D., has a great deal of experience developing and implementing a range of multi-level evaluations across sectors, routinely partnering to do so. At the time of writing, she served as the co-chair of the Evaluation Working Group of the Saskatoon Regional Intersectoral Committee, representing the Saskatchewan Federal Council and the Public Health Agency of Canada.

Isobel M. Findlay, Ph.D., is a professor in the Edwards School of Business, the university co-director of CUISR, and a scholar in the Centre for the Study of Cooperatives, University of Saskatchewan. She publishes and presents widely on Aboriginal cooperatives; partnerships, participation, and democratic governance; communications, cultures, and communities; corporate social responsibility; and transcultural law and justice.

Mary Hampton, Ph.D., is a professor of psychology at Luther College, University of Regina. In her current research, she and her community psychology research team are focusing on developing materials to increase cross-cultural knowledge and facilitate delivery of culturally appropriate end-of-life care for Aboriginal families.

Bonnie Jeffery, Ph.D., a professor of social work at the University of Regina based at the Prince Albert campus, has extensive experience conducting community based research with rural and northern communities with a focus on population health interventions to address health inequities. She is the past director and currently a faculty researcher with SPHERU, a bi-university unit of the University of Saskatchewan and the University of Regina.

Darlene Juschka, Ph.D., is an associate professor of women's and gender studies and religious studies at the University of Regina. Some of her more recent publications include *Political Bodies, Body Politic: The Semiotics of Gender* and "The Amazon, Rhetor, and Priest: Feminism, Politics, and Religion."

Wendee Kubik, Ph.D., is currently teaching at Brock University in women's studies. Previously, she was an associate professor and the coordinator of women's and gender studies at the University of Regina. Her research interests focus on farm women, women's health, Aboriginal women, gender analysis, participatory action research, food security, and violence against women.

Kristjana Loptson, M.A., is currently working on a Ph.D. in political science at the University of Alberta. As a former employee at SPHERU, she was involved in a program evaluation as well as a built environment research project. Her areas of interest include media studies, public policy, and social movement theory.

Fleur Macqueen Smith, M.A., has eighteen years of experience as a writer and editor and was the knowledge transfer manager of SPHERU's Healthy Children Research Team from 2004 to 2013. She is now the senior advisor, communications, at the Saskatchewan Advocate for Children and Youth. In 2011, she was awarded a Knowledge Translation Graduate Award by the National Collaborating Centres for Public Health for her master's research on knowledge translation and communities of practice.

Diane Martz, Ph.D., is the director of the Research Ethics Office and a research faculty member of SPHERU at the University of Saskatchewan. Her current research focuses on rural seniors' health; rural youth, risk behaviours, and healthy lifestyles; the impact of BSE on the farm family and changing rural governance structures; and research ethics.

Bev McBeth, M.Ed., RN, is an adviser with the Native Access Program to Nursing/Medicine. She has been a champion of Aboriginal nursing students and communities for inclusive and ethical research and practice.

David McDine, B.A., M.A. (candidate), was CUISR's research and communication assistant from February to August 2010. He was responsible for assisting with a variety of research projects and communication reports. He is currently working on a graduate degree at Carleton University.

Nazeem Muhajarine, Ph.D., is a social epidemiologist, the chair of Community Health and Epidemiology at the University of Saskatchewan, and director of the Saskatchewan Population Health and Evaluation Research Unit (SPHERU), where he also leads the Healthy Children Research Team. He was awarded a CIHR Knowledge Translation Award in 2006 for local/regional impact and the Saskatchewan Health Research Foundation's Achievement Award in 2009.

Clifford Ray is from Sandy Bay. From a long line of trappers and natural resource managers, he is in his third term as president of the Northern Saskatchewan Trappers Association Cooperative. Recently retired from his position with SaskPower, Clifford now puts his energy into retelling the stories of trapping and remapping the territory, including the area around Sandy Bay.

Andrea Redman is a member of Standing Buffalo First Nation with a strong cultural commitment. She served as the interim community researcher for the project carried out on Standing Buffalo.

Chief Roger Redman is the current chief at Standing Buffalo First Nation and has been a champion of the youth in his community.

Hongxia Shan, Ph.D., graduated from the University of Toronto in 2009. She is currently an assistant professor in the Department of Educational Studies of the University of British Columbia. Her research interests include immigration, lifelong learning, workplace/professional learning, community development, and qualitative methodologies.

Len Usiskin is the manager of Quint Development Corporation, a not-for-profit community economic development organization. Quint works to enhance the economic and social well-being of Saskatoon's core neighbourhoods. Len is on the board of Station 20 West Development Corporation and that of CUISR. He has an M.A. in economics from the University of Manitoba.

Janice Victor, at the time of writing, is finishing her Ph.D. in culture and human development in the Department of Psychology at the University of Saskatchewan. She plans to teach and conduct community-based research in cultural psychology following completion of her degree.

Leanne Yuzicappi is a community researcher from Standing Buffalo First Nation. She has provided leadership and support to the team and youth involved in the project conducted on Standing Buffalo.

INDEX

A
Abella, R., 36, 46
Abonyi, S., xxx
Aboriginal peoples: consultation with, 43, 82; and control of research, 36; culture of, 22, 30, 35, 42; knowledge of, 36-37; and poverty, 31; in research, 22, 24, 39, 41, 58n3, 78, 80. *See also* marginalized peoples
Aboriginal rights, 34, 36, 43
Aboriginal Tenure in the Constitution of Canada, 47
Aboriginal youth: involved in research, 146
academic partners (in research), xvii, xxii, 3, 5, 16, 140, 144, 149, 152, 163. *See also* researchers, academic
Action and Knowledge: Breaking the Monopoly with Participatory Action Research, 47
action-learning, 91-92, 97, 102
action-oriented research, 92, 94, 96, 165
advocacy: and action research, 64, 66, 68-69, 102, 156, 164, 167; and CBR, xviii, xxi, xxiv, 54, 98; for change, 56-57, 67, 92, 100-101
Advocacy for Healthy Public Policy as a Health Promotion Technology, 71
Advocacy—The Sound of Citizens' Voices: A Position Paper from the Advocacy Working Group, 72

Affinity Credit Union, 79
Alkin, M.C., 99, 105
Amara, N., 154, 161
Anderson, C., 150, 160
Archibald, L., 31, 47
Astin, A.W., 108, 119
Avalos, J., 108, 119
Avila, M.M., 130, 134
Avis, K., 150, 160

B
Bacsu, J., xxvi, 135, 166-68
Bailey, S.J., 122, 135
Baker, Q., 130, 134
Bamber, P., 108-9, 119
Bandura, A., 128n3, 134
Banerjee, S., 31, 46
Bare Bones and Feathers, 47
Barer, M.L., 161
Barker, D., xxii, xxix
Basualdo, M., xxvi, xxviii, 106
Battiste, M., 35-37, 39, 46, 49
Beauvais, C., 150, 160
Beavon, D., 48
Becker, A.B., 4, 14, 55, 71, 123, 135
Behavioural Ethics Board, 137
Bell, L., 37, 46
Belmont Report (1979), 4
Benatti, S., 109, 121
Bennett, L., 124, 135
Benson, M.J., 33, 37, 47

Bezanson, K., 78, 83, 87
Bhabha, H.K., 31, 46
Blondeau, Lori, 73n1
Blouin, D.D., 109, 119
Blum, K., 71
Bolitho, A., 13-14
Bollig, N., 121
Boris, E., 56, 70
Boser, S., 34, 46
Botes, L., 122, 134
Botkin, J.R., xxx, 14
Bourdieu, P., 38, 46
Bourner, T., 108-9, 120
Bowen, Angela, 134
Bowie, W.R., 70, 135
Boyd, S., 56-57, 64-66, 69-70
Bradbury, H., 54-55, 71, 123, 136
Bradley, B., 109, 120
Brennan, M.A., 58, 70
Bretherton, I., 128n1, 134
Brief, E., 4, 14
Bringle, R.G., 107, 109-10, 117-19
Brock, E., 165, 169
Broten, Chris, 103
Brown, B.A., 48, 55-56, 120
Brown, L.D., 72, 122, 134
Bryan, K., 124, 136
Brydon-Miller, M., 134
Bryk, A.S., 91, 105
Bubolz, M.M., 128n2, 134
Burt, R.S., 82, 87

C

Calleson, D., xxiv, xxix
Calliou, B.C., 46
Calverley, D., 40, 46
Canada's Social Economy, 88
Canadian Institute for Health Information, 160
Canadian Institutes of Health Research (CIHR), xix, xxiii, 5, 13-14, 147, 153, 158, 160, 165
Canadian Population Health Initiative (CPHI), 125, 160

capacity building, xviii, 44, 92, 100, 124
Cappello, J., 33, 35, 49
Cardinal, H., 42, 46
Carlisle, S., xxi, xxix
Carroll, W.K., 34, 46, 72, 78, 87
Castellano, M.B., 30-31, 35, 47
Castleden, H., 147
Central Urban Metis Federation Inc., 77n3, 93
Centre for the Study of Cooperatives, xx, 73n1, 74n
Champagne, F., 161
Chantier de l'économie sociale, 74, 87
Chávez, V., 130, 134
Chavis, D.M., 101, 105
CHEP Good Food Inc., 79
Chopin, N., xxviii, 164, 167-68
Chrisman, N.J., 136
Chrisp, T., 165, 169
Christopher, S., 147
CIHR Guidelines for Health Research Involving Aboriginal Peoples, 13-14
Ciske, S.J., 136
Clarke, L., xxviii, 45
Clayton, P.H., 107-8, 119
Coburn, D., xviii, xxx
Cochran, P.A., 147
Cognitive Justice in a Global World: Prudent Knowledges for a Decent Life, 49
Coleman, J.S., 30, 47, 81, 86-87
collaboration: in action learning, 92; among partners, 94, 97-98; benefits of, 154; in CBR, xxii, 6, 30, 61, 63-65, 67, 91, 124, 135, 145; CUISR-SRIC, 91, 93-96, 98-102; with marginalized peoples, 56; skills of, 108, 112
collaborative inquiry, 57, 81
Collins, R., 87
Combining Service and Learning in Higher Education: Summary Report, 120
Committee for Survivors of Torture, 62
communications, 11, 119, 126-27, 133

Index 179

Communities for Children, 151-53
Community Based Participatory Research for Health, 134
Community Campus Partnerships for Health, 167
community development, 58, 82-83, 85, 112-13, 123
Community Development Advisory Committee, 58
Community Development and Partnerships Directorate, 152, 160
community economic development, 75-76
Community Economic Development (CED), 79
Community Engaged Scholar Discussion Group, xxiii
community engagement, xxvi-xxvii, 29, 92
Community Mapping for Children in Saskatoon, 162
community partners (in research): and academic researchers, 15, 81, 84-85, 98, 113, 149, 152, 163; benefits of, 27, 112, 119; challenges of, 8, 75, 110-11; and communications, 114-16; contributions of, 17, 118, 142, 156; and CSL, 107, 109; and interviewing, 24; roles of, 117. See also researchers, community
community service learning (CSL), xxii, 107-11, 116-18
community-based participatory action research (CBPAR): as change-oriented, 56, 164; as collaborative, 55, 64, 112, 122; and CSL, 107, 110-11; and evaluation, 130; practice of, 58-59, 62, 123; process of, 54, 67, 106; projects, 64-66, 70
community-based participatory research (CBPR), 3, 53, 124, 133. See also community-based research (CBR)
Community-based Participatory Research for Health from Process to Outcomes, 71-72, 135-36

community-based research (CBR): 3 Rs of, 29-30, 36-38, 44, 86; academic support for, 116-18, 166-67; and advocacy for change, 54; as community development process, 79, 85; and CSL, 109; as democratic engagement, 73, 82-83, 85; ethical issues in, xvii, xxi-xxii, xxvi, 45; futures for, 146-47; as participatory action, xxi, 81, 102; and partnerships, xxv, 93, 98, 138, 144, 160; practice of, xvii, xxii, xxix, 96, 111, 113-14, 140, 153, 155, 163; projects, xvii, xxvi, 73, 85, 142-43, 149, 153, 158; qualities of, 145, 159; tenets of, 91-92, 100-102; value of, 168. See also community-engaged research
community-engaged research: as advocacy tool, 9, 12; as component of CSL, 108; ethical guidelines in, 13-14; as partnerships, 3, 5, 8, 117, 167; projects, 163-64, 168; risks of, 9-11. See also community-based research (CBR)
Community-University Institute for Social Research (CUISR): and communications, 114; and CSL, 107, 110-11, 118; and ethical research practice, 4-5, 29-31; as institutionary intermediary, 82; internships of, xvii-xviii, xx-xxi, xxiii, xxvi, xxix, 112-13; mandate of, 98; and partnerships, 3, 32, 74n, 91-94, 96-97, 115-17, 119, 165, 168; and social change, 106
Community-University Institute for Social Research: Partnering to Build Capacity and Connections in the Community, 121
community-university partnerships: benefits of, 166; challenges of, 85-86, 96; and community development, 83; and community engagement, 30; ethical implications of, 165; new generation of, 168; and research, xxiii-xxv, xxviii, 106, 111-12, 115, 149, 159

community-university relationships. *See* community-university partnerships
Community-University Research Alliance (CURA), xx, 14, 92, 106, 110, 165
community-university social capital, 82
Community-Up Approach, 36
Comprehensive Framework for Community Service Learning in Canada, A, 119
confidentiality, xxii, xxvi, 4, 10, 13, 17-18, 21, 23, 166
Cook, D., 147
Coombes, B., 31, 48
Cooperative Development Initiative (CDI) Innovation and Research Grant, 40-42, 45
Co-operative Membership and Globalization: New Directions in Research and Practice, 49, 87
Cooperatives Secretariat Cooperative Initiative Innovation and Research Program, 46
Core Neighbourhood Development Council (CNDC), 77n3, 78, 78n, 79
Core Neighbourhood Youth Cooperative, 79
Cornwall, A., 124, 134, 165, 169
Cotter, A., 40, 46
Cousins, J.B., 91-92, 99-100, 105
Cowan, B.G., 148
Cree people, 32
Critical Strategies for Social Research, 46, 72
Cropper, S., xxi, xxix
Crotty, M., 34, 47, 55, 70
CU Expo, xx
CUISR-SRIC collaboration. *See* collaboration
culture of agitational conversation, 80
Curtis, L.J., 150, 161
Cushon, J., 150, 160
Cutforth, N., 124, 135

D

Daes, Erica Irene, 31
Dallaire, B., 56, 71
Daniel, M., 70, 135
Darnell, C.Z., 109, 120
Davidson-Hunt, I.J., 30, 47
Davies, H., 154, 162
Decolonizing Methodologies: Research and Indigenous Peoples, 49, 169
Degagne, M., 31, 47
Delanoy, Sue, xxix, 151, 151n1, 152, 157, 162
DeLemos, J.L., 34, 47
Dene people, 32
Denis, J.L., 154-55, 161-62
Denny, K., xxx
Denzin, N.K., 48-49, 71, 126, 134
DeSantis, G., xxvii, 163-64, 166, 168
Diamantopoulos, M., xxvii, 73n1, 74n, 75, 78n, 79, 84, 87, 164, 166
Dickinson, H., 155, 161
Dimitriadis, G., 68, 71
Dirkx, J.M., 108-9, 119
Dobrohoczki, Robert, 73n1
Doing Participatory Research: A Feminist Approach, 71
Domestic Violence Unit at Family Services (Regina), 28
Drew, S.E., 139, 148
Duncan, R.E., 139, 148
Dunn, J.R., 150, 161
Duran, B., xxii, xxiv, xxxi, 55, 57, 72, 123, 130, 134, 136
Duty to Consult: New Relationships with Aboriginal Peoples, The, 48
Dyck, C., 162

E

Early Childhood Development Unit (ECDU), 125, 127, 134
Early Development Instrument (EDI), 151, 157-58

Early Development Instrument: A Population-based Measure for Communities (EDI), 161
Ebbesen, L., xxviii, 163-64, 166-68
Edwards, B., 107-8, 120
Elliott, P., 92, 100, 105
Ellis, J., 162
empowerment, 165-66, 168
End of Work: The Decline of the Global Labor Force and the Dawn of the Post-Market Era, The, 88
Eng, E., 123, 135
Engaged Scholar Day, xxiii
Engaging Contradictions: Theory, Politics, and Methods of Activist Scholarship, 169
Engaging with External Partners. Recommended Principles, Guidelines, and Action Plan Components, xxiii, xxxi
England, P., 87
ethical principles, 5, 8, 13, 168. *See also* community-engaged research; Community-University Institute for Social Research (CUISR); community-university partnerships
Evaluation and Education: At Quarter Century, 105
Evaluation Directorate Strategic Policy and Research Branch, 153, 161
Evaluation Report: Snapshot of Collaborative Processes, 105
Evaluation Working Group (EWG), xxviii, 94-96, 99-103, 105
Evans, R.G., 161
Evitts, T., 158, 162
Experience, Research, Social Change: Methods Beyond the Mainstream, 71

F

Fadem, P., 71
Fairbairn, B., 49, 74, 86-87
Fals-Borda, O., 33-34, 47
Farganis, J., 57, 70

Fay, J., 67, 70
Findlay, I.M., xxvi, 29, 31, 33, 35, 37-39, 46-48, 74n, 75, 78n, 79, 84, 87, 106, 164, 166-67
Findlay, L.M., 46
First Ministers' Early Childhood Development Agreement, 125
First Nations and Metis Consultation Policy Framework, 47
First Nations' Laws, 42
First Nations peoples. *See* Aboriginal peoples
First Nations University of Canada, 17, 137, 145
Flicker, S., xxi, xxx, 4, 14, 100-101, 105
Focus Groups: A Practical Guide for Applied Research, 71
Focus Groups as Qualitative Research, 71
Formative Evaluation of the Understanding the Early Years Initiative—June 2009, 161
Foundational Document on Outreach and Engagement: Linking with Communities for Discovery and Learning, xxiii, xxxi
Foundations of Social Research: Meaning and Perspective in the Research Process, The, 47, 70
Framing Our Directions 2010-12: Social Services and Humanities Research Council of Canada, 121
Frankish, C.J., 70, 135
Fraser, N., 56, 70
Freire, P., 57, 70, 82, 87, 123
From Problems to Strength: Appreciative Inquiry and Community Development, 87
Fuller, D.S., 109, 120
Full-time Kindergarten in Saskatchewan Part Two: An Evaluation of Full-time Kindergarten Programs in Three School Divisions, 162
Fulton, M.E., 86-87
Furco, A., 107, 119

G

Garaway, G.B., 99, 105
Garcia-Downing, C., 147
Garvin, T., 147
Gates, Rob, 134
Gaventa, J., 123, 134
Gehlert, S., xxx, 14
Gemmel, L.J., 107-8, 119
George, Deb, 28
George, M.A., 70, 135
Giles, D.E., 108-10, 117-18, 120
Gillam, L., 13-14
Giroux, H., 85, 87
Glacken, Jody, 134, 158, 162
Goodwill, Ken (Elder), 146-47
Gordon, C., 57, 70
government partners (in research), 129-32. See also policy-makers as research partners
Grabb, E.G., 56, 70
Graham, I., 161
Gray, M.J., 108-10, 116, 120
Greater Saskatoon Catholic Schools, 103
Greaves, L., 57, 71
Green, Kathryn, 134
Green, L.W., 55, 70, 122-23, 135
Greenway, M.T., 62, 71
Grover, M.S., 147
Growing Pains: Social Enterprise in Saskatoon's Core Neighbourhoods: A Case Study, 87
Guba, E., 55, 57, 62n5, 71
Guelph, University of, 167
Guidelines and Procedures: Completing a Research Participants Funding Requisition, 169
Guillemin, M., 13-14
Guillen, M.F., 87
Guta, R., 4, 14

H

Hale, C.R., 169
Halfe, L.B., 37, 47
Hall, B.L., 109, 120, 123, 134-35
Halla, W., 40, 46
Hamilton, A.C., 31, 47
Hammond Ketilson, Lou, 46, 74n
Handbook of Action Research: Concise Paperback Edition, The, 136
Handbook of Public Sociology, 87
Handbook of Qualitative Research, 134
Hankin, L., 108-9, 119
Hanson, Bev, 103
Hardy, Julia, 134
Hartsook, B., 157, 162
Hartsook, L., 157, 162
Hatcher, J.A., 107, 109-10, 117-19
Hatry, H., 62, 71
Healing Journey project, 15-18, 23, 25, 28
Health Disparity in Saskatoon: Analysis to Intervention, 105
Healthier Societies: From Analysis to Action, 161
Healthy Children research theme, xix, 150, 152, 158
Healthy Disparity in Saskatoon: Analysis to Intervention, 87
Heller, K., 101, 105
Henderson, J. Y., 33, 37, 43, 46-47
Henrika, Maria, 28
Herbert, C.P., 70, 135
Hergenrather, K.C., 148
Hertzman, C., 161
Heymann, J., 161
Hickey, C.G., 32, 41, 48
Hidden Curriculum in Higher Education, The, 48
Hidden in Plain Sight: Contributions of Aboriginal Peoples to Canadian Identity and Culture, 48
Highway, Florence, 43
Hildebrandt, W., 42, 46
Hilgendorf, A., xxxi
History of the Canadian West to 1870-1871, 48
Hofman, N.G., 109-10, 121
Hohenadel, J., 161
Holden, Bill, 45

Hondagneau-Sotelo, P., 107, 120
Horn, M., 158, 162
Howard, J.P.F., 107, 110, 120
Hrynkiw, Crandall, 103
Hubbard, P., 31, 48
Human Resources and Skills Development Canada, 151, 160-61
Human Services Integration Forum, 93
Hurley, J., 161
Huu-ay-aht First Nation, 147
Hyde, Dorothy, 103
Hyland, S., 120
Hynie, M., 108-9, 120

I

Ibáñez-Carrasco, F., xxiii- xxiv, xxx, 167, 169
Immigrant Serving Interagency Network, 63
Improving Research Dissemination and Uptake in the Health Sector: Beyond the Sound of One Hand Clapping, 161
inclusion/exclusion (in research), 55, 61, 64, 66, 70, 164
Indians in the Fur Trade, 48
Indigenous Diplomacy and the Rights of Peoples: Achieving UN Recognition, 47
Indigenous Methodologies: Characteristics, Conversations, and Contexts, 48
Indigenous peoples. *See* Aboriginal peoples
Innovations in Knowledge Translations: The SPHERU KT Casebook, 135
Institute of Aboriginal Peoples' Health (IAPH), 147
Institutional Planning and Assessment. 2nd Integrated Plan: Frequently Asked Questions, 169
Institutional Review Board, 13
Interdisciplinarity and the Transformation of the University, 87
International Institute for Sustainable Development, 81, 87

internships. *See* Community-University Institute for Social Research (CUISR) internships
interviewers, 17-19, 21-25, 27
Interviewing the Interviewers, 17, 27-28
Israel, B.A., 4, 14, 55, 71, 123-24, 135

J

Jackson, T., 134
Janus, M., 151, 161
Janzen, B., 150, 160
Jeffery, Bonnie, xix, xxx, 134
Jeffries, V., 87
Jensen, K., 120
Jenson, J., 150, 160
Jewkes, R., 124, 134
Johnny, M., 120
Jones, L., 122, 135
Jordan, C., xxiv, xxix, 35
Jordan, S., 47
Justice Interruptus: Critical Reflections on the "Postsocialist" Condition, 70

K

Kamberelis, G., 68, 71
Kendall, E., 147
Kesby, M., 35, 38, 47
Khanlou, N., 10, 14
KidsFirst Evaluation Team, 158, 162
KidsFirst program, xxviii, 122-23, 125-29, 132, 134, 160
KidsFirst Program Manual, 136
kidsKAN (www.kidskan.ca), 157-58, 160
Kiely, R., 108-10, 121
Kincheloe, J., 34, 48, 56, 71
Kindig, D., xviii, xxx
King, D.C., 107, 120
Kingsley, E., 48, 120
Kirby, S., 57, 65, 67, 71
Kleidman, R., 80, 87
knowledge, 55, 69, 92, 100, 147, 157
knowledge production, 57, 59, 64, 67-68, 70, 86, 92, 100, 108, 155

knowledge transfer, 115, 153, 155
knowledge translation, 92, 100-101, 117, 147, 149, 152-53, 159-60
Knowledge Translation in Health Care: Moving from Evidence to Practice, 161
Knows His Gun McCormick, A., 147
Kolenda, B., xxi, xxx, 100-101, 105
Kone, A., 136
Kost, R., xxx, 14
Kovach, M., 30, 37-39, 48
Krieger, J.W., 136
Krueger, R., 61n4, 71
KSI Research International Inc., 157, 161
Kubik, W., 165-66, 168
Kunst, A., 161
Kurji, A., 161
Kushner, S., 130, 135

L

Labonte, R., xviii, xxx, 56, 71, 122, 136, 162
Lamari, M., 154, 161
Lambert-Pennington, K., 120
Landry, R., 154, 161
Lavis, J., 154-55, 161-62
Lee, S.X., 48, 120
Leggett, Taban, 134
Lemieux-Charles, L., 161
Lemstra, M., 75, 77, 87, 94, 105, 150, 161
Lessans Geller, S., 56, 72
Letiecq, B.L., 122, 135
Lincoln, Y.S., 48-49, 55, 57, 62n5, 71, 126, 134
Linking, Learning, Leveraging: Social Enterprises, Knowledgeable Economies, and Sustainable Communities, 45-46
Livant, Bill, 77n2
Location of Culture, The, 46
Logic of Practice, The, 46
Lomas, J., 154-55, 159, 161
Looking Forward, Looking Back, 48
Loptson, K., xxviii
Lorde, A., 37, 48

Losing the Game: Wildlife Conservation and the Regulation of First Nations Hunting in Alberta, 1880-1930, 46
Loup, A., xxx, 14
Love, R., xxx
Loving, K., 121
Lunn, Jillian, 134
Luther College, 28

M

Mabry, J.B., 108, 120
Macdonald, S., 165, 169
Macduff, N., 53, 71
Mac-Lean, R.T., 148
Macqueen Smith, Fleur, 134-35, 152, 156, 161-64, 166-68
Maddux, H.C., 109-10, 116, 120
Maguire, P., 55, 71
Making Equality: History of Advocacy and Persons with Disabilities in Canada, 72
Manitoba, University of, 15-16
marginalized peoples, 55-57, 61n4, 68n7, 70, 84, 97, 101
Margolis, E., 34, 48
Markus, G.B., 107-8, 120
Marshall, C.A., 147
Marshall, S., 48, 120
Martin, A., xxxi
Martz, D., xxvi, 166-68
Marullo, S., 109, 121
Maslany, G., xxx
Mass Weigert, K., 109, 121
McCubbin, L., 147
McCubbin, M., 56, 71
McDine, D., xxviii
McDonald, Bart, 42
McDonald, S., 4, 14
McDonough, P., xxx
McGrath, M., xxi, xxx
McIntosh, Thomas, xix, xxx, 134
McLaren, P., 34, 48, 56, 71
McLaughlin, M.W., 105
McMaster Research Centre for the Promotion of Women's Health, 57n1

McMillan, D.W., 101, 105
McMullin, Kathleen, 134
Meager, A., 4, 14
Measuring Program Outcomes: A Practical Approach, 71
Mercer, S.L., 122, 135
Method in Social Science: A Realist Approach, 72
Methods in Community-based Participatory Research for Health, 135
Métis peoples, 32, 58n3, 78
Meyer, M., 87
Mildenberger, M., xxi, xxx, 100-101, 105
Millican, J., 108-9, 120
Mills, C., 35, 48
Ministry of Advanced Education (Saskatchewan), 103
Ministry of Education, Employment and Immigration (Saskatchewan), 103
Ministry of Social Services (Saskatchewan), 103
Minkler, M., 54, 59, 64, 71-72, 123-24, 134-36
MITACS-Accelerate, 160
Moellenbeck, Wendy, 134
Mooney, L.A., 107-8, 120
Moore, L., 71
Moote, M.A., 30, 48, 110, 120
Morgan, D., 61n4, 62n5, 71
Morton, A.S., 42, 48
Mosher-Williams, R., 56, 70
Muhajarine, Nazeem: author citations, xix, xxviii, xxix, xxx, 122, 124, 129, 135-36, 156-57, 160-62; as researcher, 125, 150-52, 158
Myers-Lipton, S.J., 108, 120
Mykhalovskiy, E., xxx

N

Natcher, D.C., 32, 41, 48
National Collaborating Centres for Public Health Knowledge Translation Awards, 162
National Longitudinal Survey of Children and Youth Community Study, 157
Native Community Care Program, 62
Natural Resources Transfer Agreement, 43
Natural Science and Engineering Council of Canada, 14
Naturalistic Inquiry, 71
Neamtan, N., 78n4, 87
Negotiating the Numbered Treaties: An Intellectual and Political Biography of William Morris, 49
Neighbourhood Profiles, 87
Nellis, M., xxxi
Nelson, M., 32, 41, 48
Nelson, R.M., xxx, 14
Netting, E., 53, 71
Neudorf, C., 75, 77, 87, 94, 105, 150, 161
New Directions in Community-based Research, xxx, 14, 105, 169
New Economic Sociology: Developments in an Emerging Field, The, 87
Newhouse, D.R., 48
Newman, D., 32, 48
Nickel, Darren, 134-35, 156, 161
Njoh, A.J., 122, 135
Non-Profit America: A Force for Democracy, 72
non-profit organizations, 56, 58-59, 61-63, 68-69
Norris-Tirrell, D., 108-9, 120
Northern and Aboriginal Health research theme, xix
Northern Ontario, Manitoba, and Saskatchewan Regional Node of the Social Economy Suite, 45-46
Northern Saskatchewan Trappers Association Cooperative (NSTAC), xxvi, 29-33, 40-41, 44, 46
Nutley, S., 154, 162
Nyden, P., 108-10, 117, 120

O

O'Connor, M., 57-58, 61, 71
Offord, D., 151, 161

Offord Centre for Child Studies, 151
Offord's Early Development Instrument, 156
O'Flaherty, R.M., 30, 47
Ondaatje, E.H., 108, 116, 120
Opondo, J., 150, 161
Outside in the Teaching Machine, 49
Overview of the Understanding the Early Years Initiative, 160
Ownership, Control, Access, and Possession (OCAP) or Self-determination Applied to Research, 14
Ozirney, Fred, 103

P

Panelli, R., 31, 48
Park, P., 134
Parker, E.A., 4, 14, 55, 71, 123, 135
Parker-Gwin, R., 107-8, 120
participation in research, 55, 57, 64, 67, 70, 92, 100, 145, 163, 165
participatory action research, 41, 92, 100, 167
Participatory Action Research: Challenges, Complications, and Opportunities, 105
participatory evaluation: as congruent with CBR, 91, 101; as inclusive, 129; as process, xxviii, 92, 122-24, 126, 130, 132; as transformative, 99-100
Partnering to Build Capacity and Connections in the Community, 49
Passelac-Ross, M.M., 43, 48
Paths to Living Well for On-Reserve Aboriginal Youth, 138
Pattison, D., 31, 38, 40, 48
Patton, M.Q., 124, 135
Pedagogy of the Oppressed, 70, 87
Perry, E.M., 109, 119
Perry, M., 71
Peter, E., 10, 14
Petrucka, P.L., xxviii, 163-64, 166, 168
Phillips, D.C., 105
Phipps, D., 120

photovoice project, 139, 141, 143
Planet S magazine, 83
Plantz, M., 62, 71
PL^{2}A^{3}Y program, 146
policy-makers as research partners, 126-27, 129, 133, 138, 150, 154. *See also* government partners (in research)
Ponting, J.R., 29, 48
Poor-Bashing: The Politics of Exclusion, 88
Poudrier, J., 148
power, 55-57, 64, 67-68, 68n7
Power/Knowledge: Selected Interviews and Other Writings 1972-1977 by Michel Foucault, 70
Prairie Action Foundation, 28
Prediger, Garry, 103
program evaluation, 124, 129
Program Managers' Committee, 126-27, 131-32
Promise and Potential: The Third Integrated Plan 2012-2016, xxxi
Proudfoot, S., 77, 87
Public Policy and Program Evaluation, 136
Puchala, C., 150, 162
Pula, S., 148
Puma, J., 124, 135
Pushor, D., 158, 162

Q

Qualitative Research and Evaluation Methods, 135
Qualitative Research in Education: Focus and Methods, 87
Quarter, J., 74, 88
Quint Development Cooperation, 74n, 78, 78n, 79

R

Race, Space, and the Law: Unmapping a White Settler Society, 48
Race and Epistemologies of Ignorance, 49
Racial Contract, The, 48

Radius Community Centre, 103
Rahman, M.A., 33, 47, 55, 71
Rain, Doug, 103
Randall, J.E., xviii, 122, 136
Raskoff, S., 107, 120
Ratner, R.S., 78, 87
Ray, A.J, 42, 48, 106
Ray, Clifford, xxvi, 32-33
Razack, S., 35, 48
Readings in Social Theory: The Classic Tradition to Post-Modernism, 70
Reardon, K.M., 107-9, 117, 120
Reason, P., 54-55, 71, 123, 136
Reddy, M., 121
Redistributing Health: New Directions in Population Health Research in Canada, xxx
reflexivity, xxv, xxvii, 29, 35, 37-39, 44, 102-3, 106-7, 164
Regina, University of, xviii, xix, xxiii-xxiv, xxxi, 17, 28, 137, 165-67, 169
Reid, C., 4, 14, 55, 57, 67, 70-72
Reitsma-Street, M., 55-56, 72
Rektor, L., 56, 72
Report of the Aboriginal Justice Inquiry of Manitoba, 47
research, as decolonizing, xxvi, 37, 39, 44, 168
Research Advisory Committee, 58
Research and Education for Solutions to Violence and Abuse (RESOLVE), 15
research ethics, 36. *See also* ethical principles
Research Ethics Board, 13, 23
research findings, ownership of, 114, 117, 124, 128, 132
Research Is Ceremony: Indigenous Research Methods, 49
researchers, academic: challenges of, 8, 11-12, 84, 122, 164; in community-engaged research, 9, 15, 17, 24, 137, 139, 154; and ethics, 168; as inclusive, 6. *See also* academic partners (in research)

researchers, community: as advocates, 9; challenges of, 7, 11-12, 164; and research process, 17, 139, 150, 154, 168. *See also* community partners (in research)
Restructuring the Relationship, 48
retention, of research participants, 25
Rethinking Critical Theory and Qualitative Research, 71
Rhodes, S.D., 148
Riaño-Alcalá, xxiii-xxiv, xxx, 167, 169
RIC Self-Assessment Tool (RSAT), 95, 100, 102
Rifkin, J., 74, 88
River Gathering Festival in Pelican Narrows, 33
Robertson, A., xxx
Rocha, C.J., xxi, xxiii, xxx
Roche, B., xxiii-xxiv, xxx, 11, 14, 100, 105, 167, 169
Rodriguez, C., 154, 162
Rogers, J., 71
Rogge, M.E., xxi, xxiii, xxx
Rosenbluth, David, 134
Rosenthal, D., 13-14
Rosing, H., 109-10, 121
Ross, L.F., xxx, 3-4, 7, 9-10, 14
Ross, N., 150, 161
Ross, S., 154, 161-62
Royal Commission on Aboriginal Peoples (RCAP), 37, 43, 48
Rrohlich, K.L., 150, 161
Rural Health research theme, xix
Russell, Gail, 134-35
Russell, N., 49, 74, 87

S

safe networking, 18
Sage Handbook of Action Research: Participative Inquiry and Practice, The, 71
Sage Handbook of Qualitative Research, The, 48-49, 71
Salamon, L., 56, 72
Salée, D., 39, 49

188 Index

Salmon, A., xxiv, xxx
Sanderson, K., 30, 49, 110, 121
Sandmann, L.R., xxii, xxx
Sanmartin, C., 150, 161
Santos, B. de, 42, 49
Sari, Nazmi, 134
Saskatchewan, Government of, 32, 43, 46-47, 93, 160; First Nations and Metis Relations, 46
Saskatchewan, University of, xviii, xix-xx, xxiv, xxxi, 74n, 92, 116, 118, 137, 160, 165-67, 169; and community-university partnerships, xxiii; Publications Fund, xxix; Research Ethics Office, 165
Saskatchewan Arts Board, 147
Saskatchewan Education, Health, Intergovernmental, and Aboriginal Affairs and Social Services, 125, 136
Saskatchewan Health Research Foundation, xix
Saskatchewan KidsFirst Program Evaluation: Summary of Findings and Recommendations, 162
Saskatchewan Knowledge to Action Network, 158
Saskatchewan Native Theatre Company, 79
Saskatchewan Population Health and Evaluation Research Unit (SPHERU), xvii-xix, xxiii, xxvi, xxix, 3-5, 53, 147, 150, 165, 168
Saskatchewan Regional Intersectoral Committee (SRIC), 91
Saskatoon, City of, xx, 74, 87, 92-93
Saskatoon Health Region, xx, 92-94
Saskatoon Policy Service, 93
Saskatoon Public Library, 157
Saskatoon Regional Intersectoral Committee (SRIC), xxviii, 92-95, 97-98, 103-5
Saskatoon Tribal Council, 78n
Saskatoon Tribal Council Urban Services Inc., 93

Saunders, M.A., 109, 121
Savan, B., xxi-xxiii, xxx, 100-101, 105
Sawyer, S.M., 139, 148
Sax, L.J., 108, 119
Sayer, A., 57, 72
Schnarch, B., 4, 14, 36, 38-39, 49
Schulz, A.J., 4, 14, 55, 71, 123, 135
Seblonka, K., xxxi
Second integrated plan: Toward an Engaged University, 2008-12, xxxi
Seifer, S.D., xxiv, xxix
Self-Determination in Action: The Entrepreneurship of the Northern Saskatchewan Trappers Association Cooperative, 48
Senturia, K.D., 136
Shan, H., xxviii, 156, 161, 163-64, 166, 168
sharing circles, 137
Shephard, Gary, 134
Sherman, R.R., 87
Shore, N., 12, 14
SHOwED format, 139
Shrader, E., 109-10, 116-17, 121
Sider, D., xxii, xxx
Siefer, S.D., 107-9, 121
Silka, L., 39, 49, 82, 88
Sinclair, C.M., 31, 47
Sister Outsider: Essays and Speeches, 48
Skulmoski, Murray, 134
Smith, D.E., 35, 49, 167, 169
Smith, F.M., xxix
Smith, G.R., xxx, 14
Smith, Jeffrey, 160
Smith, Karen, 134
Smith, L., 29-30, 35-36, 42, 49, 168-69
Smith, P., 124, 136
Smythe, D.W., 80, 88
social capital, 30, 78, 81-83, 86
social change, xxiv-xxv, xxviii, 12, 30, 66-67, 106
social economy, 74, 74n, 75-78, 78n4, 79, 81-82
social justice, 12, 35, 100

Social Sciences and Humanities
 Research Council of Canada (SSHRC),
 14-15, 32, 45, 74n, 106, 110, 119, 121,
 165, 167; Community-University
 Research Alliance Program, xx, 28
Sontag, M.S., 128n2, 134
Spingett, J., 122, 124, 136
Spivak, G.C., 37, 49
SRIC-CUISR collaboration, 91, 93-96,
 98-102
Stakeholder-based Evaluation, 105
Standing Buffalo First Nation, 137-39,
 144, 146-47
Stanton, T., xxii, xxx
Statistics Canada, 157
Stein, P., 124, 135
Stewart, P., 165, 169
Stienstra, D., 56, 72
Stoddart, G., xviii, xxx, 161
Stoecker, R., xxxi, 80, 88, 110, 121, 165,
 169
Straus, S., 161
*Study of Participatory Research
 in Health Promotion: Review
 and Recommendations for the
 Development of Participatory
 Research in Health Promotion in
 Canada*, 70, 135
Suchet-Pearson, S., 31, 48
Sullivan, M., 122, 136
Sullivan, S., 35, 49
Sultana, F., 39, 49
Swanson, J., 76, 88
Swords, A.C.S., 108-10, 121

T
Talbot, R.J., 33, 49
Tamara's House (Saskatoon), 28
Tan, B., 150, 160
Tan, L., 150, 160
Tang, S.S.-T., 166, 169
Task Force on Community-Based
 Research, xxiii
Taylor, Laura, 28

Tedmanson, D., 31, 46
Tetroe, J., 161
*Texts, Facts, and Femininity: Exploring
 the Relations of Ruling*, 49, 169
Theories of Social Inequality, 70
Third Integrated Plan: Areas of Focus,
 xxxi
Tombari, C., 124, 135
Tournier, C., 161
Toye, J., 161
Transition House (Regina), 28
trapping, 30-34, 42-44
*Trapping Rights of Aboriginal Peoples in
 Northern Alberta, The*, 48
Travers, R., 4, 14
Treaties 6, 8 and 10, 31
*Treaty Elders of Saskatchewan: Our
 Dream Is That Our Peoples Will
 One Day Be Clearly Recognized as
 Nations*, 46
treaty rights, 34, 43
*Tri-Council Policy Statement: Ethical
 Conduct for Research Involving
 Humans (TCPS)*, 4-5, 14
trust building, 14, 22, 34, 81, 83, 124, 146
Tuana, N., 35, 49
Tupper, J.A., 33, 35, 49
Turnbull, Hayley, 134
*2010 Saskatchewan Hunters' and
 Trappers' Guide*, 43, 47
Tyron, E., xxiii, xxxi

U
Understanding the Early Years (UEY)
 project, 149, 151-60
United Way of Saskatoon and Area, xxi,
 92
Ursel, Jane, 15
*Using Knowledge and Evidence in Health
 Care: Multidisciplinary Perspectives*,
 161
Usiskin, L., xxvii, 45, 74n, 164, 166
*Utilization-focused Evaluation: The New
 Century Text*, 135

V

Van Dinh, T., 80, 88
van Houten, T., 62, 71
van Rensburg, D., 122, 134
Vedung, E., 129, 136
Vega, W., 122, 134
Venne, S., 31, 49
Victor, J., xxviii
visible minority groups, 58n3, 59
voice (in CBR), 55, 57, 64, 67, 70
Voices of Change: Participatory Research in the United States and Canada, 135
Voth, D.E., 48, 120
Voyageur, C.J., 30, 48
Vu, L.T.H., 150, 160, 162

W

Walker, G.B., 48, 120
Wallerstein, N.B., xxii, xxiv, xxxi, 54-57, 67, 70-72, 122-24, 130, 134-36
Walter, I., 154, 162
Warren, J., 79, 84, 88
Waskahat, Peter, 42
Watts, V., 147
Waygood, K., xviii
Webb, R.B., 87
Weber-Pillwax, Cora, 36
Wedlock, J., 120
Weir, W., 35, 49
Well-being of Children: Are There "Neighbourhood Effects"?, The, 160
Wells, K., 122, 135
Whitmore, E., 91-92, 99-100, 105
Wiessner, S., 35-36, 39, 49
Wight-Felske, A., 56, 72
Williams, A., 122, 124, 136
Williams, K., 57-58, 61, 71
Wilson, S., 30, 36-37, 40, 49
women, abused, xxvi, 15-18, 23, 27
Wood, J., 78n, 88
Woods, Meghan, 28
Woodside, J., 154, 162
Woodward, C., 161
Woodward, G.B., xxx

Working Together towards Common Goals: Serving through Research, xxxi, 169
Workshop on Theories and Methods of Participatory Research, A, 71
Wounds of Exclusion: Poverty, Women's Health, and Social Justice, The, 72
Wright, B.D., 109, 120
Wright, J., 150, 162
Wuttunee, W., 29, 35, 39, 47, 49

Y

Young, S., 147

Z

Zakaras, L., 108, 116, 120